建筑立场系列丛书 No.29

C3

终结的建筑
Death
and Architecture

中文版

韩国C3出版公社 | 编

于风军 陶源 刘愁琼 郑江婷 徐雨晨 | 译

大连理工大学出版社

终结的建筑

004 终结的建筑 _ *Silvio Carta*

建造终端建筑

006 建造终端建筑 _ *Silvio Carta*

010 De Nieuwe Ooster 墓园 _ Karres en Brands Landscape Architects

024 莱克伍德花园公墓 _ HGA

042 帕尔马火葬场 _ Zermani Associati Studio di Architettura

056 阿尔塔奇的伊斯兰墓园 _ Architekt di Bernardo Bader

066 新希尔维公墓 _ Giovanni Vaccarini

072 云神殿 _ Clavel Arquitectos

084 茵格海姆葬礼小教堂 _ Bayer & Strobel Architekten

094 埃伦巴赫公墓访客中心 _ Andreas Fuhrimann Gabrielle Hächler Architekten

102 Kedainiai 第一火葬场 _ G.Natkevicius & Partners

108 皮诺索殡仪馆和花园 _ Cor & Asociados

纪念性建筑

118 纪念性建筑：设计某种缺失的元素 _ *Nelson Mota*

122 废奴纪念馆 _ Wodiczko + Bonder

136 911国家纪念馆 _ Handel Architects

144 墨西哥暴乱遇难者纪念馆 _ Gaeta-Springall Arquitectos

154 集中营纪念馆 _ awg Architecten

166 Zanis Lipke纪念馆 _ Zaigas Gailes Birojs

176 帕尔米拉纪念博物馆 _ WXCA

004 *Death and Architecture_Silvio Carta*

Building the End

006 *A Space for the End_Silvio Carta*

010 De Nieuwe Ooster Cemetery Park_Karres en Brands Landscape Architects

024 Lakewood Garden Mausoleum_HGA

042 Parma Crematory_Zermani Associati Studio di Architettura

056 Islamic Cemetery in Altach_Architekt di Bernardo Bader

066 New Silvi Cemetery_Giovanni Vaccarini

072 Cloud Pantheon_Clavel Arquitectos

084 Ingelheim Funeral Chapel_Bayer & Strobel Architekten

094 Erlenbach Cemetery Building_Andreas Fuhrimann Gabrielle Hächler Architekten

102 The First Crematorium in Kedainiai_G.Natkevicius & Partners

108 Pinoso Funeral Home and Garden_Cor & Asociados

Architecture of Memorial

118 *Architecture of Memorial: Designing the Presence of Something Absent_Nelson Mota*

122 Memorial to the Abolition of Slavery_Wodiczko + Bonder

136 National September 11 Memorial_Handel Architects

144 Memorial to Victims of Violence in Mexico_Gaeta-Springall Arquitectos

154 Kazerne Dossin_awg Architecten

166 Zanis Lipke Memorial_Zaigas Gailes Birojs

176 Memorial Museum in Palmiry_WXCA

终结的建筑

Death Archit[ecture]

本期C3主要介绍建筑遗迹。建筑是由人类所建造，且反映了它们比人类本身的存在时间长的这一事实。然而，同一人在其一生中，可能会被根本不存在现实世界的人所预先"构建"，而他也可能会享受其所处时代的"构建"过程。在这种程度上，建筑成为一种周期性存在的连接网，将不同辈分，年代甚至是不同世纪的人们交织在一起，作为对人类生活的一个沉默的见证。

尽管建筑的周期性作用可能并不经常是其最重要的一个方面，但是它却更多地嵌入在建筑的内在特点中。一些建筑还带有特殊的建造目标，以记录人类生命消亡的时刻。

建筑类型是多种多样的，并且反映了不同的文化以及思考生命终结的方式。

本文阐述了一个较为明确的分化：殡仪性建筑，即"建造终端建筑"，包括了与死亡相关的，为一切事物提供便利和实施场所的建筑，以及纪念性建筑，即"纪念建筑"，它们都可以作为从生存过渡到死亡的地标性建筑。从建筑中延伸出，一方面致力于展现终结的时刻，另一方面则纪念试图展现永恒的人。

这两种终结建筑具有几个共同点，同时也有不同之处。其中最主要的一个不同点可能是——当殡仪性建筑置于带有宗教性质的人类遗迹

This issue of C3 deals with the architecture of the aftermath. Architecture is built by people with the projection that buildings will last longer than their individual existence. Nonetheless the same individuals – during their life – fully enjoy the buildings of their period, which – in turn – may have been previously built by those who are no longer in the physical world. By this extent, architecture becomes the connecting tissue of a cyclical existence, intertwining individuals from different generations, decades and even centuries, and it stays as a silent witness of people's life.

Although maybe the cyclical role of architecture is not always the most prominent of architecture's aspect, but it stays more embedded within its intrinsic characteristics, some projects are built with the specific goal of addressing the moment of people's end. The typology is quite varied and reflects different cultures and ways of considering the end.

A clear differentiation is made in this issue: funerary architectures *Building the End*, encompassing buildings which facilitate and house all the chores related to death, and memorial architectures *Architecture of Memorial*, intended as buildings landmarking the survival over death. The two sections stretch from the buildings dedicated to the moment of the end, to those who try to render the eternity.

The two "architectures for the end" share several aspects and – at the same time – differ for many others. One main distinction may be that – while the funerary architecture is the place allocated to religiously store people's remains, that is crematoria, columbaria,

and
ecture

中，即火葬场、骨灰堂、葬礼教堂、停尸房或地下墓室——纪念碑则成为庆祝生命的地方，且为想拥抱共享回忆的活人而建。

第一组项目描述了试图战胜空间的终端性的空间设计方法，包含了一系列不同的建筑特点（根据其所在的文化生态圈，7页）。在建立这种终端性时，建筑通过侧面影像（能够帮助用户建立战胜死亡的感觉，7页）来展示项目。而由侧面影像所产生的叙述性描述允许人们在"景观、内部元素的连接以及互相冲突的两个世界之间的临界处"来表达建筑（7页）。第一批入选的十个项目似乎通过一个"驱走悲伤的支柱"系在一起，即一种迷失感或者人类无法完全理解某件事情之前存在的一种简单的不正确性（9页）。

另一方面。第二批项目在时间透视图中描绘了人类的悲伤。"在许多纪念建筑中——Nelson Mota在他的文章中阐述到——最令人吃惊的印象就是出现了个人的印记，如名字或者是可唤起回忆的数字，以作为一个代表性的介质，抑或是表达悲伤的一种方式（119页）

而在"纪念性建筑"这一章节所展示的项目中，描述了建筑师在纪念性建筑中处理"个人经历"的不同方法。

本书中所有的项目可以被看成是一个整体，它们表达了人类的悲伤和希望，以及我们置于建筑形式中的强烈信念，以应对这种情感和复杂的人性。

funerary chapels, morgues or crypts – memorials are common places of celebration of life, intended for living persons that want to embrace a common remembrance. The first group *"describes various approaches to try to make space overcome the 'end', and embody a variety of architectural characteristics depending on the cultural ecology in which they sit"*. (p.7) In *Building the end* the projects are presented through the lens of a side image *"which can help users of the building overcome the sense of death"*. (p.7) The narrative generated by the side image allows to articulate the projects in *"landscape, articulation of internal elements and threshold between two worlds set one against the other"*. (p.7) The ten projects of the first section seem to be tied together by a "brace against grief", that is *"the sense of being lost or simply of inadequacy before something is ultimately not comprehensible to the human mind"*. (p.9)

On the other hand, the second section portrays the human grief in the temporal perspective. *"In many memorials"* – argues Nelson Mota in his text – *"one of the most striking phenomena is the use of tokens of the individual, such as names or evocative numbers, as a representational medium, an expression of grief"*. (p.119) The projects presented in *Architecture of Memorial* describes a variety of ways in which architects have dealt with the "individual experience" in memorials. Considered as a whole, the projects of this issue speak about human grief and human hope, as well as the strong belief we all put in architectural forms in order to cope with such feelings and the complex understanding of human nature. Silvio Carta

建造终端建筑

Building the End

关于建筑的几个定义都包含了如下问题:怎样满足人类的要求,怎样在其内举办活动、展示力量、缅怀过去,或者是——从最近的来说——怎样引发全新的且更复杂的社会规划。所有的案例都意味着可以为活着的人们传递建筑的意义和功能。但是,当我们需要建筑能够应付生命的终结时,将会发生什么样的事情呢? 用于举办葬礼的建筑对于展现人类存在的两个阶段(即生命与死亡)之间的临界空间来说,发挥着微妙的作用。这种建筑与其他建筑类型不同,它们需要创造一种宗教氛围,在这里,人们对面积、灯光、颜色以及材料均有所要求,以为对死亡表达的信念、缅怀、悲伤以及哀悼创造一个合适的平台。C3之前曾在主题为"缅怀形式"的第323期(原版)以及"圣坛永不落"的312期和其他期中也讨论过相似的问题。然而,本期所分析的项目将会阐述这类特殊建筑作品的一个重要方面,即不断地且普遍地展示人类生命的开始和结束,自然元素对于建筑师来说,似乎是一个较为关键的设计元素,以面对终端建筑所带来的挑战。

我们通常根据葬礼类别的分类法来对建筑进行分类,以为人类生命的最后时刻服务。对死亡的崇拜的传统在人类历史上存在了很长时间,且从史前开始,便一直伴随着人类文明的左右。而与死亡相关的建筑的特点则取决于其所嵌入的和反映的文化(对生命与死亡之间的关系所产生的文化观点)。希望、恐惧以及信念都被具体地包含在这类特殊的建筑中。回顾这类作品的历史,人们可以在死亡的背景下表达出古老文明的信念所产生的优雅形象。例如,在一些文明中,去世的人下葬时会陪葬一些必要的物件,以用于来世的生活,使人们相信死亡只是两个世界的一个过渡过程。埃及金字塔便是应用这种理念的闻名于世界的一个典型。其他文化也有关于死亡后肉体的假设,一些文化认为当灵魂(或者思想)存在,肉体便无所用;其他则认为肉体会重生,还有些人则将死亡看做是人类个体存在的终结。所有这些关于死亡的理念都实际地反映在殡仪性建筑的形式、空间布局以及普遍性特点中。关于死亡的这一列解释清单很长,且各不相同,但是只有几篇重要文章可能展现了足够的洞察力,允许我们看到这一主题的一个有用的全景视图。法国作者Michel Ragon的《地下幽深处》[1](1981年),以及James Stevens Curl的《死亡的赞歌》(对这些建筑、纪念物、具有西方欧洲传统的殡仪性建筑的背景进行介绍),或者是Helaine Silverman的《死亡的空间场所》[2]便

Several definitions of architecture encompass such questions as how to meet human needs, accommodate activities, display power, remember past events, or – most recently – trigger new and more complex social schemes. All these examples are meant to convey meanings and functions for living people. But what happens when we require architecture to deal with the aftermath of life? Funerary architecture has the delicate duty of representing a threshold between two phases of human existence: life and its end. More than in other architectural typologies, funerary architecture needs to create a religious atmosphere in which proportions, light, colors and materials are all called upon to provide an adequate platform for the belief, remembrance, grief and tribute that follow a loss. C3 has previously discussed similar questions in The Form of Remembrance(#323) and Places of Worship Never End(#312) and elsewhere. However, the analysis of the projects in this issue will shed light on a significant aspect of these special works of architecture. Constantly and ubiquitously present at the beginning and ending of our lives, natural elements appear to be a crucial design element for architects called upon to face the challenge of building for the end.

What we normally categorize under the taxonomy of funerary typology is architecture devoted to the final moment of people's lives. The cult of death has a long tradition in human history and has from prehistory been a constant accompaniment to civilization. The characteristics of death-related architecture depend on the culture in which it is embedded and how it reflects that culture's view of the relationship between life and death. Hopes, fears, and beliefs are all concretely embodied in this special architecture. Looking back at the history of such work, one may derive an elegant image of the faith that ancient civilizations expressed in the context of death. In some civilizations, for instance, dead people were buried with necessary objects for the afterlife in the belief that death was a moment of transition between two worlds. Egyptian pyramids are one world-renowned example of this idea. Other cultures had their own hypotheses concerning the physical body after death, some believing that it becomes useless while the soul (or mind) survives, others believing in the reincarnation, and others conceiving of death as the final end of individual human existence. All these ideas of death are physically reflected in the forms, spatial organization and general architectural characteristics of funerary buildings. The list of such interpretations of death can be quite long and diverse, yet a few key texts may provide sufficient insight to allow us to derive a useful panorama of the topic. French author Michel Ragon's L'espace de la mort[1](1981), may be one such text, along with James Stevens Curl's *A celebration of death: An introduction to some of the buildings, monuments, and settings of funerary architecture in the western European tradition*, or *The Space and Place of Death* by Helaine Silverman.[2]

De Nieuwe Ooster墓园_De Nieuwe Ooster Cemetery Park/Karres en Brands Landscape Architects
莱克伍德花园公墓_Lakewood Garden Mausoleum/HGA
帕尔马火葬场_Parma Crematory/Zermani Associati Studio di Architettura
阿尔塔奇的伊斯兰墓园_Islamic Cemetery in Altach/Architekt di Bernardo Bader
新希尔维公墓_New Silvi Cemetery/Giovanni Vaccarini
云神殿_Cloud Pantheon/Clavel Arquitectos
茵格海姆葬礼小教堂_Ingelheim Funeral Chapel/Bayer & Strobel Architekten
埃伦巴赫公墓访客中心_Erlenbach Cemetery Building/Andreas Fuhrimann Gabrielle Hächler Architekten
Kedainiai第一火葬场_The First Crematorium in Kedainiai/G.Natkevicius & Partners
皮诺索殡仪馆和花园_Pinoso Funeral Home and Garden/Cor & Asociados

终结的空间_A Space for the End/Silvio Carta

是这类书籍。

然而，人们赋予这种转变时刻的注意事项，以及它们是怎样在空间与形式中反映出来的，均与我们的谈话有关。

让我们来看一下这种死亡感是怎样被翻译成建筑形式的。一个普遍的特点似乎浮出水面：无法忍受的概念。无论一种文化所秉持的死亡理念是什么，人类生命的"终结"都被广泛地感知，以至于其范围无法衡量，其真实性也无法被深度地接受。为了面对这种绝对的死亡理念，人类似乎采取了非同寻常的方法。人们可能注意到某些墓地建造纪念碑或者是殡仪馆的意图，即庆祝人类反抗生命终结的能力和力量，而这种意图也在与建筑相关的诗学中有所出现，以将观察者与其居住的某个地方的物质性隔离开来，使他/她的思想无法沁入形而上学的维度空间。为了描述第一种方法，人们可能会回想起金字塔、豪华的坟墓，或者是一些杰出的纪念物。由Mario Fiorentino, Giuseppe Perugini等人设计的，位于罗马的Fosse Ardeatine纪念碑（1952年）包含了一块混凝土巨石，看似"漂浮"在地面上，是第一组建筑的一个典范。而像César Portela的菲尼斯特雷公墓（西班牙，2000年），Giovanni Vaccarini的奥托纳公墓（意大利，2006年）以及Aldo Rossi的圣加大公墓（意大利，1984年）则是第二组项目的代表。当第一组建筑似乎是通过实现大面积或者外形较大的建筑的姿态来理解死亡理念时，第二组建筑则较为依赖一种形而上学的氛围（在这个氛围内，人类的实质存在性与非人类属性之间的界限变得模糊不清）的形成。

这种"情感支撑"也出现在用来"庆祝"死亡的仪式中。仪式通常是在一个神圣的时刻所执行的一系列预先设计好的动作，以战胜死亡。根据意大利的社会学家Ferrero[3]的观点，人类在面对不人道的事情时，倾向于围绕在某些自然事物的周围（如火），并且举行一个仪式，这意味着将其从大众的恐惧中驱赶出来。

但是，如果在一些社交活动中，这种对抗死亡的精神支撑是由仪式来提供的，那么这种供给是怎样转化为建筑形式以及具体的空间条件呢？

以下的项目描述各种尝试展示空间终结的方法，并且依据其所在的文化生态，包含了各种各样的建筑特点。我们可以通过这些规划的镜头来对这些项目进行解读，即将它们看成是一种"侧面影像"，它能帮助用户战胜死亡感[4]。这种侧面影像可以通过各种方式来进行表达：景观、室内元素或者是两个相互冲突的世界之间的临界处。

例如，在Zermani Associati Studio di Architettura设计的帕尔马火葬场中，周期景观似乎发挥了关键的作用，宽敞的室内庭院为火葬场提

However, what is relevant to our discourse are the considerations people have given to the moment of transition and how these have been reflected in space and form.
Looking at how the sense of death has been translated into architectural forms, a common characteristic seems to emerge: the notion of the unbearable. Whatever idea of death a culture may hold, the "end" of human life seems to be felt so immense that its extent cannot be measured, nor its reality deeply accepted. In order to confront the overwhelming idea of death, humankind seems to resort to the extraordinary. One may note this intention in the monumentality of certain cemeteries or funerary buildings, celebrating the power and the force of humankind against its end, or in the poetics of architectural shapes, detaching the observer from the physicality of the places where he/she resides and projecting his/her mind into a metaphysical dimension. To picture the first approach one may recall pyramids, sumptuous tombs, or outstanding memorials. The Monument of Fosse Ardeatine(1952) in Rome by Mario Fiorentino, Giuseppe Perugini et al. – which consists of a huge concrete monolith literally floating above the ground – is an eloquent example of this first group. Projects such as César Portela's Finisterre Cemetery, Spain(2000), Giovanni Vaccarini's Ortona Cemetery (2006), Italy, or Aldo Rossi's San Cataldo Cemetery (1984), Italy may represent the second group. While the first group seems to cope with the idea of death by realizing a gesture of huge proportions or shapes, the second seems to rely on the generation of a sort of metaphysical atmosphere in which the physicality of human life is blurred into the non-human.

This sort of "emotional bracing" is also present in the exercise of rituals employed to "celebrate" death. The ritual is usually a set of pre-ordered actions performed in a solemn attempt to overcome the idea of death. According to Italian protosociologist Ferrero[3], faced with the inhuman, people tend to gather around something natural (e.g. a fire) and practice a ritual meant to defend them from a common fear.
But if in social interaction this bracing against the idea of death is provided by ritual, how is this provision translated into architectural forms and concrete spatial conditions?
The following projects describe various approaches to try to make space overcome the "end", and embody a variety of architectural characteristics depending on the cultural ecology in which they sit. The proposed lens through which we may read these projects is to see all of them as generating a "side image" which can help users of the building overcome the sense of death.[4] The side image, can be articulated in various ways: as landscape, as an articulation of internal elements or as a threshold between two worlds set one against the other.
In Zermani Associati Studio di Architettura's Crematory in the city of Parma, for instance, a crucial role seems to be given to the surrounding landscape. The generous inner courtyard provides the internal spatiality of the crematorium with a large area of expansion, and – at the same time – it represents a constant extra presence. The dark, blurry, brick-clad interior space is countered by the sharp, direct natural light from outside.
Similarly, the Erlenbach Cemetery Building by Swiss firm Andreas

由Mario Fiorentino、Giuseppe Perugini等于1952年设计的、位于罗马的Fosse Ardeatine纪念碑
The Monument of Fosse Ardeatine in Rome by Mario Fiorentino, Giuseppe Perugini et al., 1952

由Giovanni Vaccarini于2006年设计的位于意大利基耶蒂省的奥托纳公墓
Ortona Cemetery in Chieti, Italy by Giovanni Vaccarini, 2006

供了内部空间性，并且还可大面积地扩建。同时，该建筑还成为一种长期的、特别存在的建筑的代表。黑色、模糊、镶有砖覆层的室内空间与强烈的、从室外射入的自然光线形成对比。类似的，由瑞士公司Andreas Fuhrimann Gabrielle Hächler Architekten设计的Erlenbach墓地建筑似乎是依据苏黎世湖北部的视野而建的，以讲述一个相对应的故事（这个故事旨在与死亡的理念如影随形）。该建筑的特点是其内部为压缩型空间，其中的每个物体都较为朴素，且呈裸露的状态，与建筑师反对的壮观的外向视野相对立。在室内，人们在几处地方可以看见各种各样的窥视孔，从天窗到彩色玻璃表面和预制混凝土墙体，它们使人们的注意力转移到周围自然全景所产生的美感中。自然再一次地利用更伟大的事物来面对悲伤。

人们可能在位于意大利泰拉莫的Giovanni Vaccarini设计的新Silvi墓地中发现通过自然景观来建造侧面影像的手法。从其清晰的压缩型空间（产生了用于沉思的室内空间以及服务于死亡的宗教氛围）来看，该项目提供了周围小山的全景视野。Vaccarini通过打开室外的混凝土墙体，来产生突发性美的一瞥，也因此在新墓地的两个主翼之间遗留出缝隙空间，抑或是大型裂口。

虽然本文讨论的这个项目的主要特点对大部分所展示的项目来说都适用，但是还是有一些方面在某些例子中占主导地位。例如，由Clavel Arquitectos设计的、位于穆尔西亚的云神殿与Cor & Asociados事务所设计的位于阿利坎特的皮诺索殡仪馆和花园（均位于西班牙）似乎将室内的对比性（黑暗和光明）进行了强化。当然，这种对比性产生了强烈的象征性内涵。

尽管是一个简单的陵墓，但是前者却通过采用与主表皮的严谨形状形成对比的不规则表面，来产生了一个复杂的室内空间连接方式。陵墓的室内照明通过主要的正旋转门的室内照明来形成双层光照系统。如果旋转门完全关闭，发出白色光线的单一主体量便成为了焦点，使室内成为黑色且未知的世界。相反，如果门是开着的，室内空间便暴露出来，允许游客看透室内空间的丰富性。建筑师将室内空间看做是一朵白色的云彩，抑或是通过半透明的玛瑙后墙以及抛光的白色大理石墙体形成的倒影"着色"所产生的光效应。

皮诺索殡仪馆和花园沿着弯曲的道路而建，其毛坯表面与大型玻璃洞口交替产生的对比也瞬时蜿蜒。光影间的互动产生了一系列空间，从令人眩晕的氛围到安抚人心的氛围。黑色喷漆外表面与白色覆层室内激发人们进行沉思，并且展示了具有象征意义的矛盾对比：光明与黑暗、白色与黑色、生命与死亡。

人们根据其他的项目则可洞察过渡空间的定义，过渡空间提供了一种陈述，可以帮助人们维持死亡的理念。位于立陶宛、由G.Natkevicius及

Fuhrimann Gabrielle Hächler Architekten seems to rely on a view of the northern part of Lake Zürich in order to generate a parallel story whose task is to accompany the idea of death. The building is characterized by a compressed spatiality on the inside, where everything tends to be sober and bare, against which the architects have opposed a spectacular outward view. From the inside, people encounter in several locations peepholes of various types, from skylights to colored glass surfaces and perforated concrete walls, which drive their attentions toward the beauty of the surrounding natural panorama. Again, the natural presence is meant to confront grief with something greater.

The construction of a side image through natural landscape is something one may also observe in Giovanni Vaccarini's New Silvi Cemetery, Teramo, Italy. From its articulated encompassed space, which already generates an inner world of meditation and a religious atmosphere dedicated to the rest of death, the project offers sudden views of the surrounding hilly panorama. Vaccarini creates glimpses of sudden beauty by opening the outer concrete walls, hence leaving interstitial spaces, or large gaps, between the two long main wings constituting the new cemetery.

Although the main characteristics here discussed are valid for almost all the projects presented, some aspects appear to predominate in certain cases. For example, the Cloud Panteon by Clavel Arquitectos in Murcia and the Pinoso Funeral Home and Garden in Alicante by Cor & Asociados, both in Spain, appear to accent the internal contrast they generate between darkness and light. Of course, this contrast acquires strong symbolic connotations.

Despite being a simple mausoleum, the former work generates a complex internal articulation of spaces by employing irregular surfaces that contrast with the rigorous shape of the main envelope. The mausoleum performs a double light play depending on internal illumination via the main front pivoting doors. If the doors are fully shut, the white-glowing main monolithic volume acquires emphasis, leaving the interior dark and unknown. By contrast, as the doors open the inside is unveiled, allowing visitors to see through the richness of an inner space which the architects envision as a white cloud, a glowing light effect rendered via the translucent rear wall of onyx and its reflection on the polished white marble.

The Pinoso Funeral Home and garden meanders along a winding course, along which sudden contrasts between blank surfaces and large glass openings alternate. The interplay of daylight and shadow generates a variety of spaces ranging from disorienting ambiences to reassuring atmospheres. Dark-painted exterior surfaces and white-clad interiors stimulate contemplation and suggest a symbolical reference to the antinomy of light and darkness, white and black, life and death.

Other projects offer insights into the notion of a transitional space that offers a narrative to help one sustain the idea of death. The First Crematorium in Kedainiai, Lithuania, designed by G.Natkevicius & Partners, hinges on the idea of creating an inner world within its own walls. The open structure of the main enve-

1. Ragon, M., *The space of death: a study of funerary architecture, decoration, and urbanism*, (Eng. tr. Alan Sheridan), University Press of Virginia, 1983
2. Silverman, H., *Introduction: The Space and Place of Death*, Archeological Papers of the American Anthropological Association, 2002
3. cf. Ferrero, G., *The Principles of Power: the Great Political Crises of History*, (Eng. tr. T.R. Jaeckel), G.P. Putman's Sons, 1942
4. It is relevant to note that the idea of death is not here presented as either positive or negative, but merely as a "fact" of human life that is greater than individuals can bear.

合伙人设计的Kedainiai第一火葬场，则取决于在围墙内创造一个内在世界的理念。尽管其主要的围护结构呈现开放的状态，且与外面相连，但是这一情况却几乎被忽略：大型混凝土墙体是预制的，与室内空间几乎没有任何联系；相反，室内的大型玻璃窗户却面向室内庭院，而不是与周围环境所相连。

同样，位于奥地利福拉尔贝格的阿尔塔奇伊斯兰墓地，由Bernardo Bader设计，这处公墓提供的过渡空间没有在室内和室外之间划分出明确的界限。木质的伊斯兰八角花纹装饰在其与外面世界的关系中发挥着中介的作用。该项目室内空间的流动性似乎是通过狭长的通道（将各个室内房间、覆顶区域以及可观看墓地区的洞口连接起来）来蜿蜒前进。整体的空间性交替地扩张与收缩，产生了一种引人冥想的室内复杂性。

De Nieuwe Ooster墓园，由Karres en Brands景观建筑师事务所设计，试图展现了一个带状的室内空间。狭长的体量产生了一个内部世界，在这个世界中，这种唯一的与外界的接触是通过顶部来自天空的视野亦或是通过将体量分成若干部分的断裂处来实现的。新室内空间将体量自身围合起来，但是却使其对公墓的其他部分处于开放的状态。

由Bayer&Strobel建筑事务所设计的茵格海姆葬礼小教堂与Kedainiai第一火葬场有些许的类似，至少与上述分析相关。设计师通过交替设置小型室内庭院，来设计一处室内的冥想空间。过渡空间既没有完全被覆盖，也非呈现完全开放或者封闭的状态。这座建筑是由一系列不同的空间构成，这些空间总是处于一种中间的状态，即位于室内和室外之间，日光与阴影之间。

这些介于由悲伤所定义的室内空间和展现永恒且毫无拘束的宏伟和希望的外部世界之间的空间结构，在美国明尼苏达州明尼阿波利斯的莱克伍德花园公墓中也有所体现。这个项目展现了人们可以在此观察到的方方面面。莱克伍德花园公墓通过围合空间以及在某些地点唤起墙外自然环境的存在，来创造一处室内过渡空间。这是一处受保护的安全地带，以回顾生命的最后一刻。但是此地却展现了自然的壮观所产生的影响，允许自然通过天窗、玻璃表面、墙体上的洞以及倒影（水池倒影和发光的大理石铺路倒影）渗透进来，同时通过对周围全景环境的一瞥得以体现。

所有的这些项目都展示了人类支撑悲伤的强烈意愿，即一种迷失感，抑或是一种在人类未完全了解某个事物之前的一种简单的不适应性。这种精神支柱在这些建筑里成形，并致力于容纳死亡的概念，无论它是否被看做是一场终结、一次过渡，还是一次解放。建筑，如同文中所展示的例子，使用了一些自然元素来战胜死亡，利用自然来与自然和处于其间的人类互动。

lope, contacted with the outside appears in this case to be almost neglected: the large concrete walls are perforated, but little visual relationship with the inner spaces exists; on the contrary, the large glass windows of the inner rooms face inner courtyards, rather than relating to the surroundings.

Similarly, the Islamic Cemetery in Altach, Vorarlberg, Austria, by Bernardo Bader, offers transitional spaces neither clearly inner or outer. The relationship with the outside world is here mediated by the wooden Islamic octagon-patterned ornamentation. The inner flow of this project's space seems to snake through the elongated passageways connecting the various inner rooms, covered areas and sudden openings over the view of the grave area. The overall spatiality alternately expands and contracts, generating a meditative inner complexity.

De Nieuwe Ooster Cemetery Park, by Karres en Brands Landscape Architects displays an attempt to create inner spaces in a long strip. The elongated volume generates an inner world in which the sole direct contact with the "outside" is via the sky view at the top or the sudden vertical breaks dividing the volume's several parts. The newly generated interior space encloses the volume itself, yet leaves it completely open to the rest of the cemetery.

Slightly similar to the crematorium in Kedainiai, at least as concerns this analysis, is the Ingelheim Funeral Chapel by Bayer & Strobel Architekten. An inner contemplative dimension is created through the alternation of small inner courtyards, intermediate spaces neither entirely covered, nor completely open or closed. The building is a sequence of varying spaces always in an "in-between" condition – between inside and outside and daylight and shade.

The construction of spaces mediating between an inside space characterized by grief, and a world outside presented as a constant and unlimited source of greatness and hope, can also be observed in the Lakewood Garden Mausoleum in Minneapolis, Minnesota(USA). The project presents aspects that one may observe in almost all those presented here. The Lakewood Mausoleum generates an inner dimension of meditation both by enclosing spaces and by evoking at certain spots the presence of the natural environment outside the walls. The building is a protected and safe place for reflecting upon life's final moments, yet it allows the influence of the greatness of nature to permeate from outside, through skylights, glass surfaces, holes in the walls, and reflections(both from water ponds and the shining marble pavements) and through glimpses of the surrounding panorama.

All these projects display a strong human need for a brace against grief, the sense of being lost or simply of inadequacy before something ultimately not comprehensible to the human mind. This bracing takes shape in those buildings dedicated to housing the notion of death, whether conceived as an end, a transition or a liberation. Architecture – as the presented examples show – strongly employs natural elements to overcome death, playing nature against nature, with humanity in between. Silvio Carta

De Nieuwe Ooster墓园

Karres en Brands Landscape Architects

墓地

墓园是社会的一面镜子，现在是，并将一直是。通过墓园，人们可以透视个人与集体的关系、当时的社会环境、整体自然景色、殡葬文化以及设计和景观建筑领域的发展。De Nieuwe Ooster墓园位于阿姆斯特丹，也是迄今为止荷兰最大的公墓（从坟墓数量上来说），其中的纪念花园由Karres en Brands景观建筑师事务所设计。De Nieuwe Ooster墓园分三个阶段修建，分别在1889年、1915年和1928年。第一阶段和第二阶段的修建由伦纳德·施普林格（Leonard Springer）设计。墓园的这些部分具有鲜明的特有的空间品质，而第三阶段却没有保持这种空间品质。尽管空间设计与施普林格的风格相似，但不尽相同，改建和扩建使其没有了以往的结构和同一性。纪念花园就属于这一阶段的设计。建筑师们没有试图从空间上连接三个区域，而是认为有必要使每个区域都有自己独立的身份特征。通过增加对比度，墓园被明确划分为三个部分，每个区域的空间品质都得到了增强。第三阶段的墓园设计拥有新的身份特征，设计师对空间进行了强有力却简洁明了的干预。设计的基础就是设计一个由宽度不等的平行带组成的区域，而每一条平行带都有自己的设计原则。在这样各有特点的设计结构中，个人可根据意愿做出选择。有的地方用树篱分割出一个个空间，现存的墓地和纪念花园融为一体，就像房间镶上了绿色的边缘，白桦树散落在整个区域中。细长的池塘和存放骨灰瓮的围墙进一步强调了空间感，这里是存放火化骨灰的特别所在地。

骨灰安置所

作为阿姆斯特丹De Nieuwe Ooster墓园火葬场纪念花园的组成部分，Karres en Brands设计了一个放置骨灰瓮的壁龛，即骨灰安置所。

骨灰瓮建筑是构成整个区域的平行条带之一，可以存放1000个骨灰盒。建筑体量狭长，几条通道将整个建筑分成几块，通过外墙和屋顶表面的斜线结构连为一体。骨灰瓮内设计了许多存放骨灰盒的房间。骨灰瓮建筑长120m，宽5m，高5m，从外面看就像是个含蓄而健壮的锌制雕塑。而建筑内部呈白色，每个房间都是封闭的，宁静祥和。房间里的访客与周围环境隔离开来，目光所及只有白色的水磨石墙壁和头顶的天空。墙壁上设计了一系列单一和双位壁龛，在墙上某些关键的地方也预留了一些特别的洞口，既可看一眼外部周围环境，又可让光线进入其内。

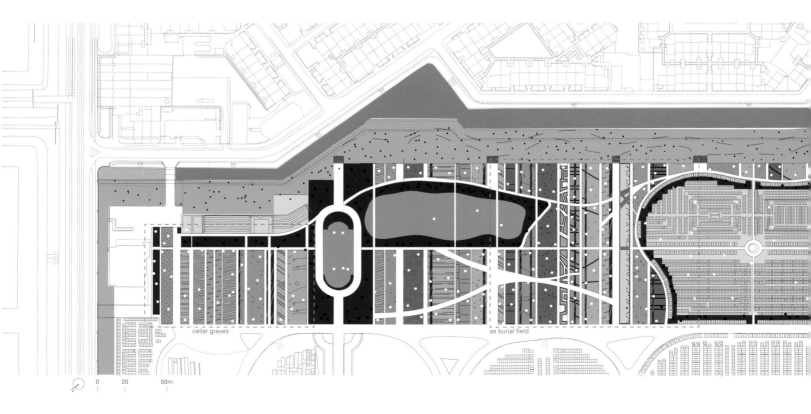

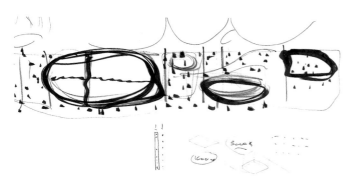

De Nieuwe Ooster Cemetery Park

As Burial Field

Cemeteries have always been, and still are, reflections of society: they provide an insight into the relationship between the collective and the individual, the social environment of the time, the overall natural scene, the funerary culture and developments in the field of design and landscape architecture. Karres en Brands created a design for the garden of remembrance of the De Nieuwe Ooster cemetery in Amsterdam, the largest cemetery (in terms of number of graves) in the Netherlands. The Nieuwe Ooster was laid out in three phases: in 1889, 1915 and 1928. The first and second phases were designed by Leonard Springer. These sections have a clear spatial quality all of their own, but the third phase does not share this quality. It bears a resemblance to the style of Springer, but is not the same. Adaptations and expansions have left it devoid of structure and identity. The garden of remembrance lies within this phase. Instead of spatially linking the three zones, the

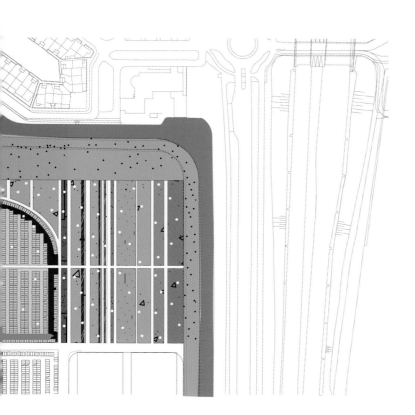

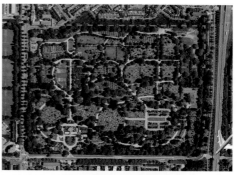

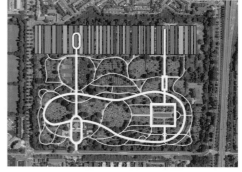

项目名称：De Nieuwe Ooster Cemetery
地点：Watergraafsmeer, Amsterdam, The Netherlands
建筑师：Karres en Brands Landscape Architects
项目团队：Sylvia Karres, Bart Brands, Lieneke van Campen, Joost de Natris, James Melsom, Alejandro Noe, Marc Springer, Jim Navarro, Julien Merle, Pierre-Alexandre Marchevet
技术规范与指导：Rod'or Advies
结构设计：Van der Toll
甲方：De Nieuwe Ooster Cemetery
用地面积：total_330,000m², as burial field_20,000m², columbarium area_15,000m²
设计时间：2005~2006(phase 1), 2009(phase 2)
竣工时间：2005~2008(phase 1), 2011(phase 2)
摄影师：courtesy of the architect-p.12 top, p.12 bottom-right, p.14, p.15, p.16, p.17, p.18, p.22 bottom
©Aerial Shooting (courtesy of the architect)-p.13 top, p.16~17
©Jeroen Musch (courtesy of the architect)-p.12 middle, p.12 bottom-left, p.13 middle, p.20, p.21, p.23 bottom (except as noted)

architects found it necessary to give each area its own separate identity. By increasing the contrast, a clear triple division of the cemetery is brought about, so that the qualities of each individual zone are enhanced. A new identity has been created for the third phase. A robust but simple intervention was called for here. The basis is a zone with parallel strips of varying widths, each with its own design principle. Within this unambiguous structure, choices are made possible for individual wishes. Some of the strips include hedges that divide the zone into spatial compartments. The existing graveyards and the garden of remembrance are incorporated into the zone like rooms with green edges. Birch trees are loosely spread throughout the zone as a whole. An elongated pond and an urn wall form spatial accents, and a special destination for cremation ashes.

Columbarium

Karres en Brands has designed a columbarium – a place for storing funeral urns – as a component of the crematorium garden of remembrance at the De Nieuwe Ooster cemetery in Amsterdam. The building forms one of the strips that structure the area, and provides places for 1,000 urns. The columbarium is an elongated volume dissected by pathways. The separate elements of the building created in this way are linked by slanted lines in the exterior walls and roof. A number of rooms to house the urns are hollowed out within the volume. The outside of the building – which is 120 meters long, 5 meters wide and 5 meters high – has the appearance of an introverted and robust zinc sculpture. From the inside, the rooms form enclosed, peaceful white interiors. Within the rooms visitors are sheltered from the surroundings, only the white terrazzo walls and the sky are visible. A pattern of single and double niches is hollowed out of the walls. At certain key points there are special niches where openings in the walls offer a glimpse of the surroundings and let in light.

一层 first floor

屋顶 roof

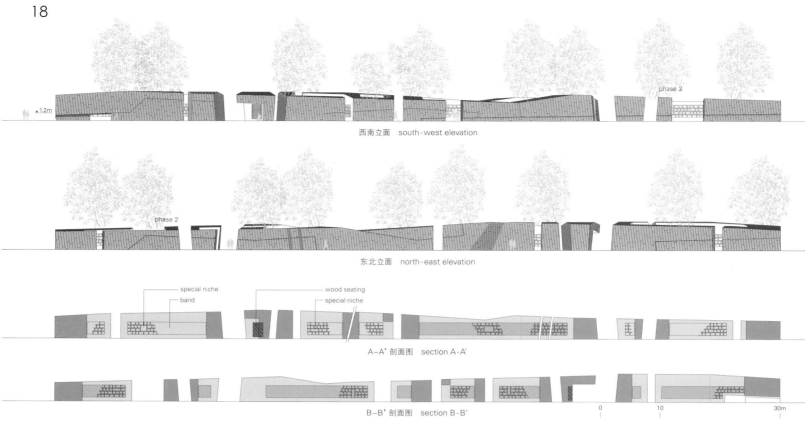

西南立面 south-west elevation

东北立面 north-east elevation

A-A' 剖面图 section A-A'

B-B' 剖面图 section B-B'

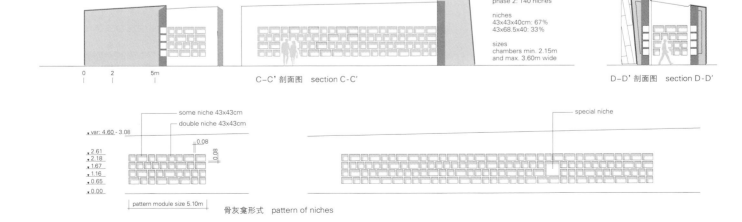

capacity
phase 1: 750 niches
phase 2: 140 niches

niches
43x43x40cm: 67%
43x68.5x40: 33%

sizes
chambers min. 2.15m
and max. 3.60m wide

C-C' 剖面图 section C-C'

D-D' 剖面图 section D-D'

骨灰龛形式 pattern of niches

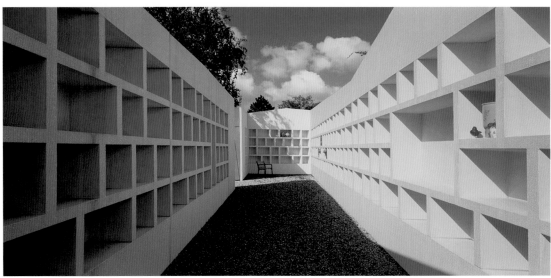

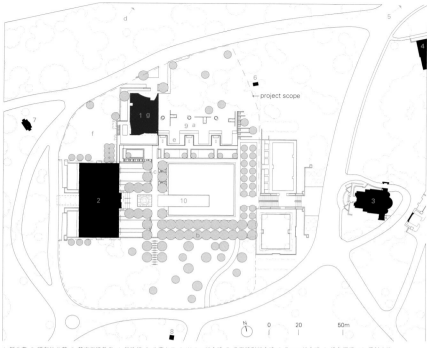

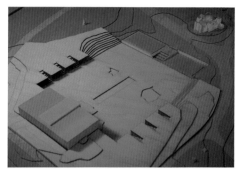

1 新公墓 2 现存的公墓 3 莱克伍德教堂 4 行政楼 5 公墓入口 6 Walker纪念碑 7 弗里德利纪念碑 8 Pence纪念碑 9 绿色屋顶 10 反射水池

a 绿色屋顶 b 场地雨水处理系统 c 小型防渗透表面 d 临近的公交和自行车路线 e 具有保温性能的地下结构 f 用于灌溉的零饮用水
g 能源效率高的建筑系统、白色屋顶、主要的零能耗材料、充足的阳光和南向场地、低流动性的设备

1. new mausoleum 2. existing mausoleum 3. lakewood chapel 4. administration building 5. cemetery entrance
6. walker monument 7. fridley monument 8. pence monument 9. green roof 10. reflecting pool
sustainability features
a. green roof b. on site stormwater management c. minimal impermeable surfaces d. adjacent bus line and bike path e. thermal benefits from subterranean structure f. zero potable water used for irrigation g. energy efficient building systems; white roof; predominance of zero emission materials; abundant daylight and south-facing site; low flow fixtures

莱克伍德花园公墓

HGA

莱克伍德花园公墓的设计灵感来自于巴黎Pere-Lachaise公墓的景观设计,但完全遵循了鲜明的美国化的草坪绿化公墓的传统:大的家族式纪念碑和个人墓碑混杂交织在一起,置于开阔的、修剪平整的草坪之上,周围由各种树木、宁静的湖泊和柔和弯曲的小路环绕。

因为现有的陵墓容量接近饱和,莱克伍德计划设计建造一处2276m²的新陵墓,包括可安葬10 000多人的安葬区、一个小教堂、接待中心和16 187m²新景观区。

在原来深受人们喜爱的地方添加一大型结构充满了挑战,因此HGA建筑事务所决定采用保护和强化公墓的历史景观来作为开发设计策略。三分之二的项目深藏于一个山坡中,使街道标高的体块最小化。底层建筑的绿色屋顶进一步延伸了原有公墓的草坪,而草丘上棱角分明的天窗为地下建筑空间提供了自然采光。在陵墓的入口,白色的马赛克图案环环相连,镶嵌在白色的墙面上,如波浪般翻滚,让人们再次联想到历史上久负盛名的1910年莱克伍德教堂错综复杂、丰富多彩的马赛克镶嵌内饰。横条纹的灰色花岗岩外墙面使整座建筑与大地融为一体。青铜大门引领访客进入一处祥和宁静的空间:桃花心木墙面、大理石地板、充足的阳光,还有将墓地和周围景色尽收眼底的开阔视野。

在入口处,人们经过数级楼梯,可以到达地下一层。往西走,一整面威尼斯风格的灰泥墙引导哀悼者通往一个有45个座位的小教堂。向东延伸,一个走廊将骨灰龛凹进处(用来放置火化后的骨灰)和地下墓室(用来放置灵柩)串为一体。虽然每个膛室具有相似的几何结构,但都有细微的设计变化。晶莹明亮的玛瑙地板或蜂蜜色,或绿色,或粉红,交替变换。窗户和天窗的方向也根据房间的位置旋转、改变。所以,所能看到的景致也各不相同,或看见近景,或远眺远处的地平线,或向上仰望树冠或晴朗的天空。

自始至终,所选材质形成的鲜明对比——颜色或浅或深,表面或粗糙或光滑,风格或纯朴或精致,无论在视觉上还是在触觉上,都为人们沉思冥想、怀念追忆提供了一处安静的空间。同时,这一设计还认识到将死亡看做生活问题,在思考死亡时,人们有不同的观点和独特的愿望。在为一万人设计一个永久安息之所时,本设计充分尊重个性、人的尺度和与自然世界感官的联系,认为这些是最重要的。

Lakewood Garden Mausoleum

Inspired by the landscape of Pere-Lachaise Cemetery in Paris, Lakewood follows the distinctly Americanized tradition of the lawn plan cemetery – a mix of large family monuments and individual grave markers arranged within open, sweeping lawns framed by masses of trees, serene lakes and softly curving roads. With an existing mausoleum nearing capacity, Lakewood developed a plan for a new 24,500sf mausoleum that includes burial space for more than 10,000 people, a chapel, reception center,

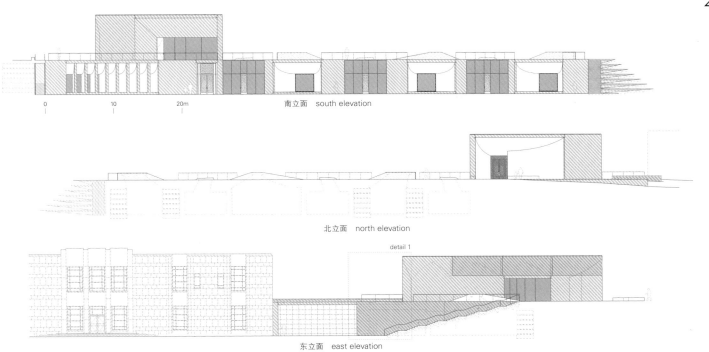

南立面 south elevation

北立面 north elevation

东立面 east elevation

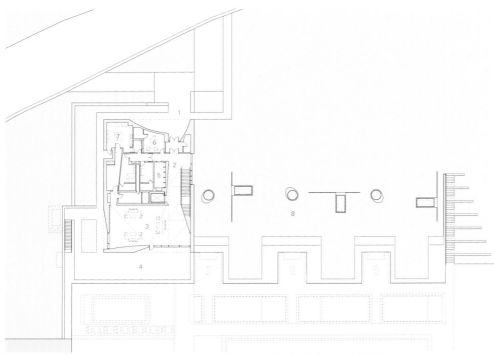

1 入口/站点 2 门厅 3 接待处 4 南侧露台 5 更衣室 6 办公室 7 餐饮厨房 8 绿色屋顶
1. entry / dropoff 2. foyer 3. reception room 4. south terrace 5. coat room 6. office 7. catering kitchen 8. green roof
二层　second floor

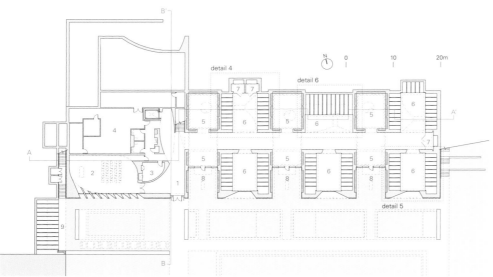

1 低层门厅 2 教堂 3 默哀室 4 机械室 5 骨灰安置室 6 地下室 7 家庭使用的地下室 8 骨灰安置花园 9 花园地下室
1. lower foyer 2. chapel 3. grieving room 4. mechanical room 5. columbarium room
6. crypt room 7. family crypt room 8. columbarium garden 9. garden crypts
一层　first floor

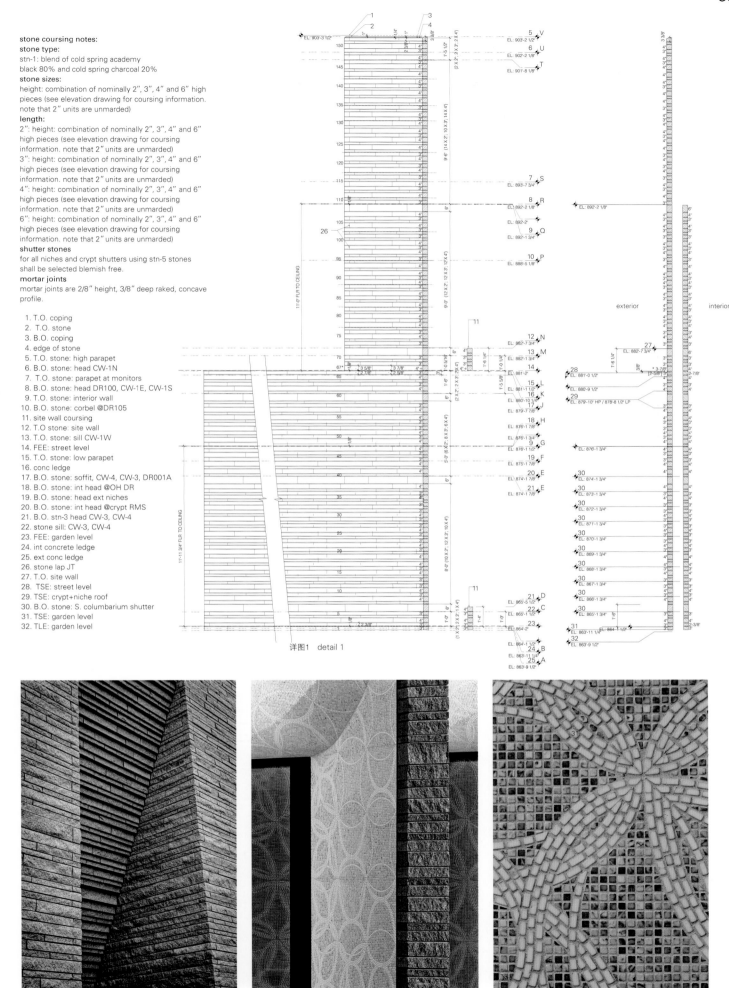

stone coursing notes:
stone type:
stn-1: blend of cold spring academy black 80% and cold spring charcoal 20%
stone sizes:
height: combination of nominally 2", 3", 4" and 6" high pieces (see elevation drawing for coursing information. note that 2" units are unmarded)
length:
2": height: combination of nominally 2", 3", 4" and 6" high pieces (see elevation drawing for coursing information. note that 2" units are unmarded)
3": height: combination of nominally 2", 3", 4" and 6" high pieces (see elevation drawing for coursing information. note that 2" units are unmarded)
4": height: combination of nominally 2", 3", 4" and 6" high pieces (see elevation drawing for coursing information. note that 2" units are unmarded)
6": height: combination of nominally 2", 3", 4" and 6" high pieces (see elevation drawing for coursing information. note that 2" units are unmarded)
shutter stones
for all niches and crypt shutters using stn-5 stones shall be selected blemish free.
mortar joints
mortar joints are 2/8" height, 3/8" deep raked, concave profile.

1. T.O. coping
2. T.O. stone
3. B.O. coping
4. edge of stone
5. T.O. stone: high parapet
6. B.O. stone: head CW-1N
7. T.O. stone: parapet at monitors
8. B.O. stone: head DR100, CW-1E, CW-1S
9. T.O. stone: interior wall
10. B.O. stone: corbel @DR105
11. site wall coursing
12. T.O stone: site wall
13. T.O. stone: sill CW-1W
14. FEE: street level
15. T.O. stone: low parapet
16. conc ledge
17. B.O. stone: soffit, CW-4, CW-3, DR001A
18. B.O. stone: int head @OH DR
19. B.O. stone: head ext niches
20. B.O. stone: int head @crypt RMS
21. B.O. stn-3 head CW-3, CW-4
22. stone sill: CW-3, CW-4
23. FEE: garden level
24. int concrete ledge
25. ext conc ledge
26. stone lap JT
27. T.O. site wall
28. TSE: street level
29. TSE: crypt+niche roof
30. B.O. stone: S. columbarium shutter
31. TSE: garden level
32. TLE: garden level

详图1 detail 1

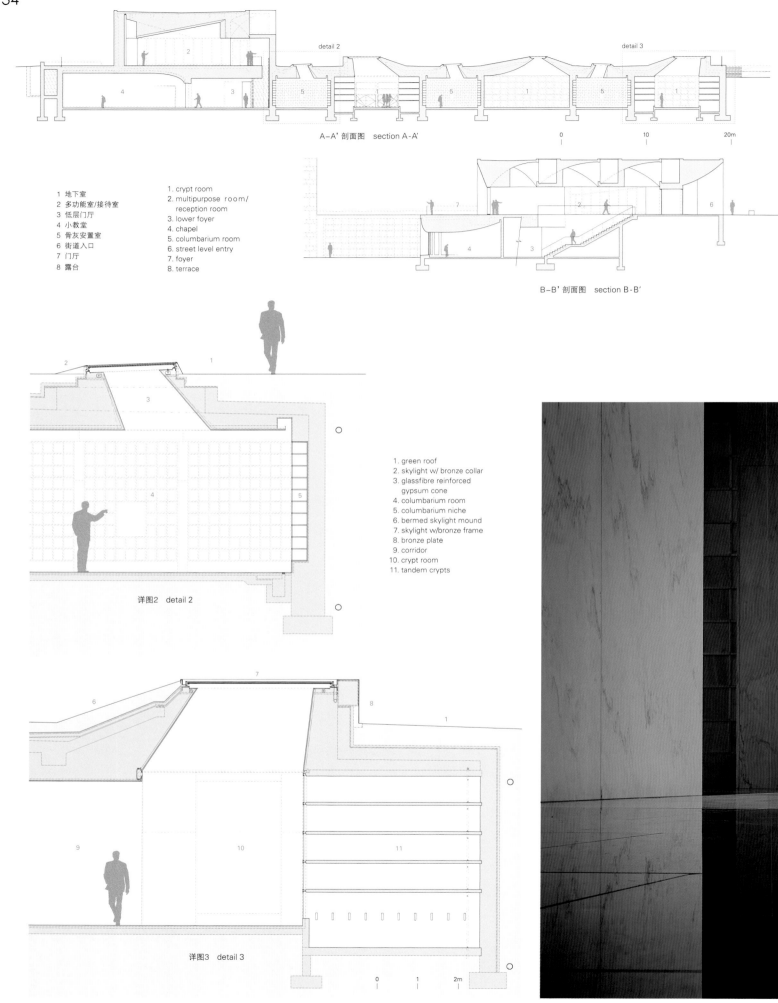

A-A' 剖面图 section A-A'

B-B' 剖面图 section B-B'

1 地下室	1. crypt room
2 多功能室/接待室	2. multipurpose room/reception room
3 低层门厅	3. lower foyer
4 小教堂	4. chapel
5 骨灰安置室	5. columbarium room
6 街道入口	6. street level entry
7 门厅	7. foyer
8 露台	8. terrace

1. green roof
2. skylight w/ bronze collar
3. glassfibre reinforced gypsum cone
4. columbarium room
5. columbarium niche
6. bermed skylight mound
7. skylight w/bronze frame
8. bronze plate
9. corridor
10. crypt room
11. tandem crypts

详图2 detail 2

详图3 detail 3

项目名称：Lakewood Cemetery Garden Mausoleum
地点：Minneapolis, Minnesota
建筑师：HGA Architects and Engineers
主要委托人：Daniel Avchen
设计委托人：Joan M. Soranno
项目经理：Stephen Fiskum
项目建筑师：John Cook
项目团队：Nick Potts, Michael Koch, Steve Philippi, Jay Lane, Ross Altheimer, Robert Johnson Miller
景观建筑师：Halvorson Design Partnership
总承包商：M.A. Mortenson Company
甲方：Lakewood Cemetery Association
用地面积：16,187m²
有效楼层面积：2,276m²
竣工时间：2012
摄影师：©Paul Crosby (courtesy of the architect)

and new landscaping on four acres.

Challenged with adding a large structure to a much-beloved place, HGA developed a strategy that protected and enhanced the cemetery's historic landscape. Two-thirds of the program is tucked into a hillside to minimize the massing at the street level. A green roof planted over the lower level extends the cemetery's lawn when angled grass mounds articulate skylights for the building's subterranean spaces. At the Mausoleum's entry, a white mosaic pattern rendered in infinite loops across white billowing surfaces reimagines the historic 1910 Lakewood Chapel's intricate and colorful mosaic tesserae interiors. Horizontal bands of split-faced gray granite tie the structure to the earth. Bronze doors usher visitors into a serene space of mahogany walls, marble floors, abundant daylight, and expansive views of the cemetery and landscape.

A generously scaled stair draws visitors from the entry to the lower garden level. To the west, a sweeping Venetian plaster wall directs mourners to a 45-seat chapel. Stretching east, a single hallway strings together alternating bays of columbarium (for cremated remains) and crypt (for caskets) rooms. While geometrically similar, each chamber is distinguished by subtle design variations. Inset floors of luminous onyx alternate between honey, green, and pink. Window and skylight orientations rotate and shift between rooms, variously framing views to near or distant horizons or up to the tree canopy or clear sky.

Throughout, the contrast of textures – light and dark, rough and smooth, rustic and refined – calls upon both visual and tactile senses to offer a peaceful place for contemplation and remembrance. The design also recognizes that in contemplating death – as in living matters – people have diverse perspectives and desire uniqueness. It respects that in designing a final resting place for ten thousand people, individuality, human scale, and a sensory connection to the natural world are paramount.

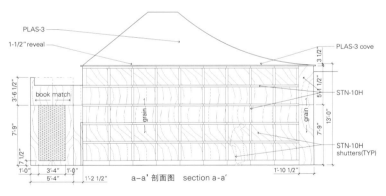

详图4 detail 4

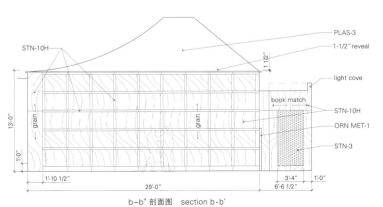

a-a' 剖面图 section a-a'

b-b' 剖面图 section b-b'

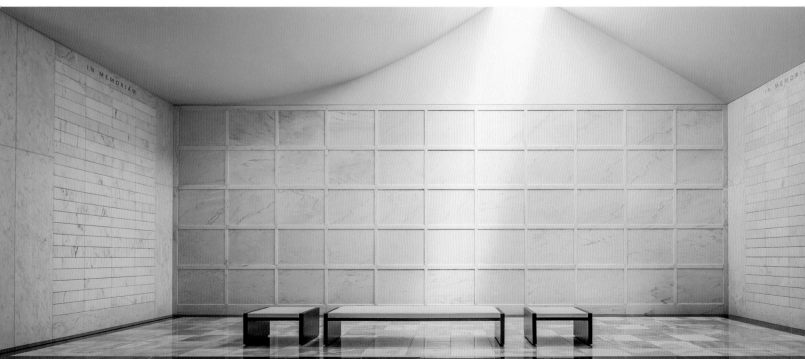

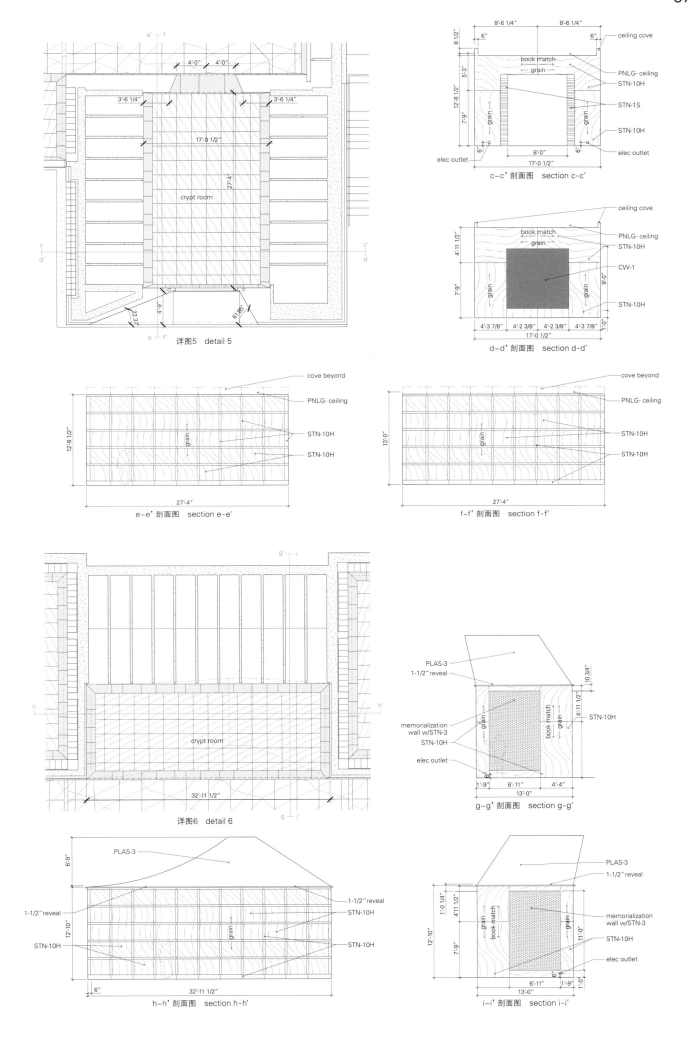

Death and Architecture Building the End

帕尔马火葬场
Zermani Associati Studio di Architettura

新建的帕尔马火葬场位于瓦勒拉公墓的北部,瓦勒拉公墓和新建的环城公路之间,距帕尔马以西大约一公里处。

该项目一边是城市和环城公路,另一边是乡村和瓦勒拉小镇,其景观设计既体现了一个世纪殖民历史的鲜明特征,也体现了基本的罗马中世纪道路:这一文明至今仍然展现在罗马住所中,展现在道路和农场的布局上,展现在Vicofertile的罗马式建筑里。

该项目所要解决的主要问题是如何处理好原有的建筑围墙、新设计的公墓围墙与其所在的乡村和瓦勒拉小镇之间的关系。

项目中柱廊状的建筑围墙是存放骨灰的地方,内部有一条连续的通道,它暗含了生与死之间的关系,可以解读为一种理想的生死轮回的连续感。

新建的帕尔马火葬场紧挨着老的公墓,是一座大型长方形建筑,从靠近原有公墓一侧的停车场可以到达。这一拱廊式建筑体现出一种时间上的层次结构,其建筑媒介就是位于这两个时间维度之间的寺庙本身。

这也从空间上定义了室内和室外举办仪式的时间顺序,将逝者到达的区域与接待逝者家人的区域分开,而这处区域在建成后则靠近入口,远离扬洒骨灰的花园。

告别大厅呈正方形,其背景墙上竖向切割开来一个开口,光线明亮,这里是将遗体运往焚化室的大门,那里将有各种技术环境对遗体进行处理。

遗体就这样消失在火光中。

十字架以其内在的仪式性定义了不同时刻的空间层次结构,这一结构连续不断地进行重建,如同走廊所体现的永恒轮回一样,将所有的一切都包裹在无穷而永恒的旅程中。

寺庙就在围墙之内,远远就可看到,那些绕过城镇中心旁道的人都能看到,如同巴西利卡教堂平面中的一个大碎片,前面加上一个大型前柱式构造,南面和北面相似,分别朝向瓦勒拉镇和帕尔马市。

整座建筑通过走廊通道的形式使时间驻足,使它成为一个大城市的象征,借此,这座城市不断地通过悼念逝者来庆祝城市自身的记忆。

Parma Crematory

The new Parma Crematory is located on the north of the Valera Cemetery, between this one and the newly built ring road, approximately 1km away from west of the city.

On one side the city and the ring road, on the other one the countryside and the town of Valera, mark the references to a landscape characterized by the centurial history of colonization and the basic Roman altomedievale roads: a civilization still readable in the Roman Domus, in the layout of roads and farms, in the Vicofertile's Romanesque architecture.

The relationship between the two fences, old and designed and that between them, the countryside and the town of Valera is the

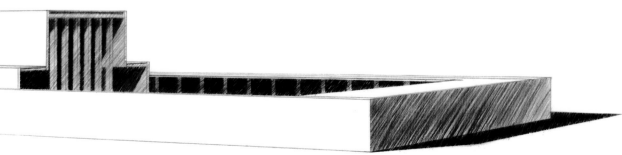

46

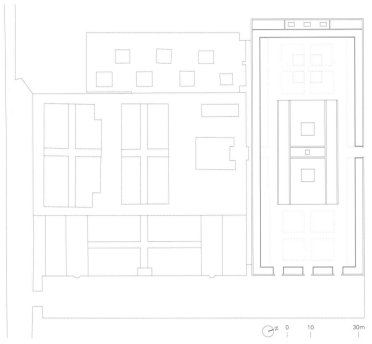

项目名称：Parma Crematory
地点：Valera, Parma, Italy
结构工程师：Paola Tanzi
设备工程师：Massimiliano Brioni
机械工程师：Corrado Ceccardi
用地面积：5,235m²
总建筑面积：1,143m²
有效楼层面积：1,032m²
竣工时间：2010
摄影师：©Mauro Davoli (courtesy of the architect)

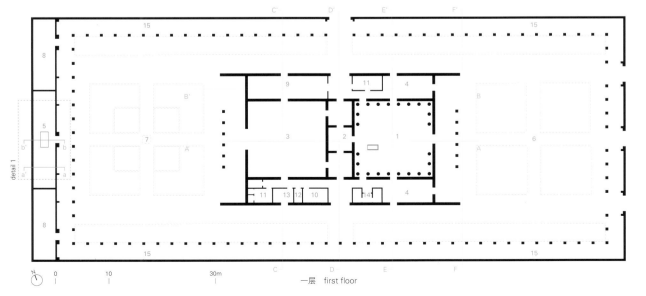

1 告别大厅
2 罗马式神殿
3 火化大厅
4 骨灰等待室
5 普通的骨灰存放室
6 前花园
7 缅怀花园
8 撒扬骨灰的花园
9 死者遗体的临时存放处
10 办公室
11 更衣室
12 骨灰瓮存放室
13 控制室
14 浴室
15 盛骨灰处

1. hall of farewell
2. sacellum
3. hall of cremation
4. waiting room for the ashes
5. common cinerary
6. front garden
7. garden of remembrance
8. garden of sprinkling
9. temporary storage for mortals rem
10. office
11. dressing room
12. storage for urns
13. control room
14. bathroom
15. cineraries

一层 first floor

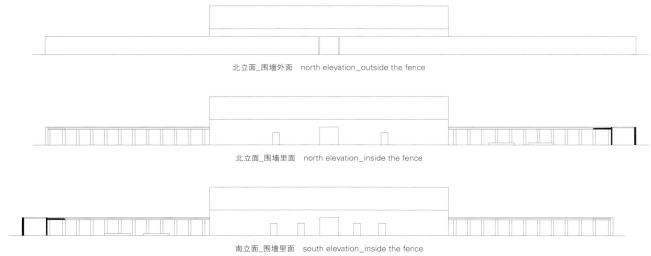

北立面_围墙外面 north elevation_outside the fence

北立面_围墙里面 north elevation_inside the fence

南立面_围墙里面 south elevation_inside the fence

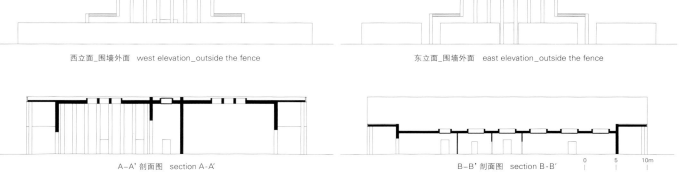

西立面_围墙外面 west elevation_outside the fence

东立面_围墙外面 east elevation_outside the fence

A-A' 剖面图 section A-A'

B-B' 剖面图 section B-B'

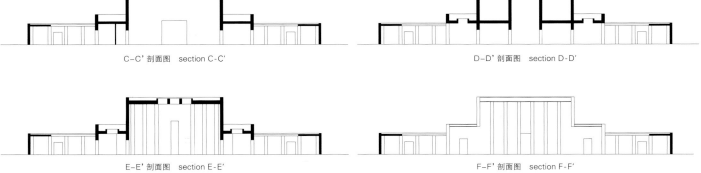

C-C' 剖面图　section C-C'　　　　　　　　　　　D-D' 剖面图　section D-D'

E-E' 剖面图　section E-E'　　　　　　　　　　　F-F' 剖面图　section F-F'

详图1 detail 1

详图1 立面 detail 1 elevation

a-a' 剖面图 section a-a'

b-b' 剖面图 section b-b'

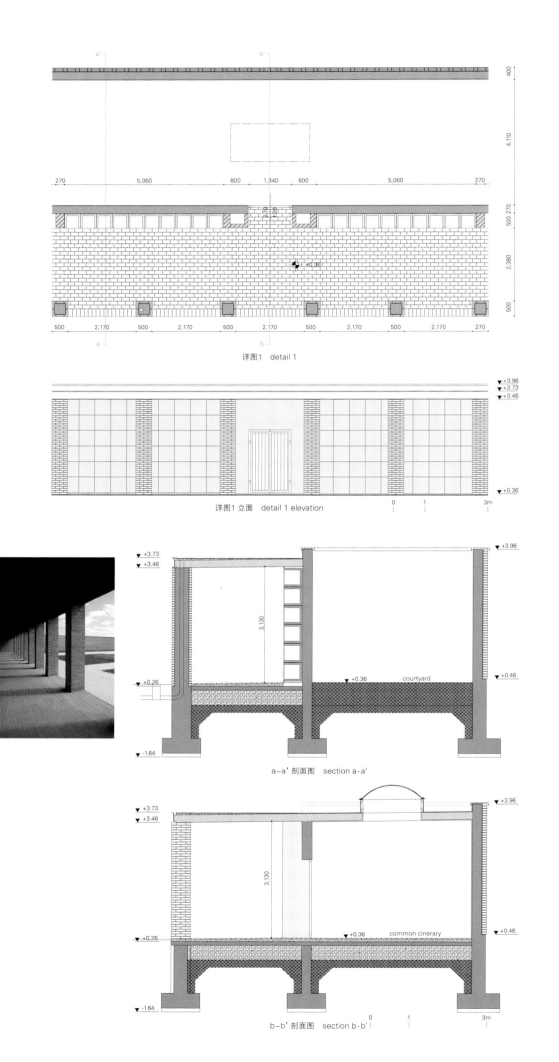

main issue addressed by the project.

The fence, was made of architectural space that designed as a porticoed wall and inhabited by cellar containing dust, contains, in an uninterrupted path, the relationship between life and death, establishing a reading in the sense of the ideal continuity.

In the form of a large rectangle the building which is next to the existing arcade-style cemetery, which can be reached from a parking lot located on the closer side, according to the present use of the existing cemetery, contains the moments of a hierarchy, whose architectural medium is the temple itself, in the middle of the two dimensions.

This also spatially marks the time of the rite, between exterior and interior, dividing the area where the deceased and their families are received, located near the entrance, from the garden of sprinkling of ashes, and after construction.

The hall of farewell, a square, is characterized by a vertical cut of the light on the back wall which is also the gate to transition of the body to a zenithal room lit and technical environments.

The body thus disappears in the light.

The crossing, with its internal rituality marks the spatial hierarchy of the different moments, which is continually reconstructed, as in a cycle of eternity from the porch, and wraps everything in an infinite and timeless journey.

The temple emerges within the fence, visible from afar and from those who use the bypass, such as a large fragment with the Basilica plan, preceded by a large prostyle similar to the south and north, towards Valera and Parma.

This piece, suspends in time the rite of the passage, making it one big urban symbol in which the city celebrates, in an incessant way, the memory of itself through the memory of its dead.

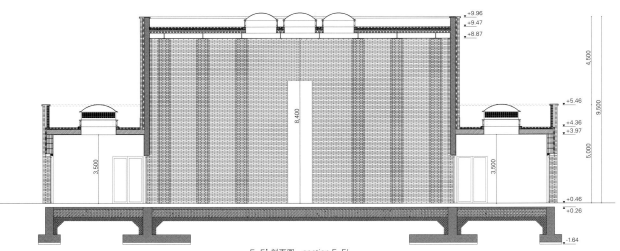

E-E' 剖面图　section E-E'

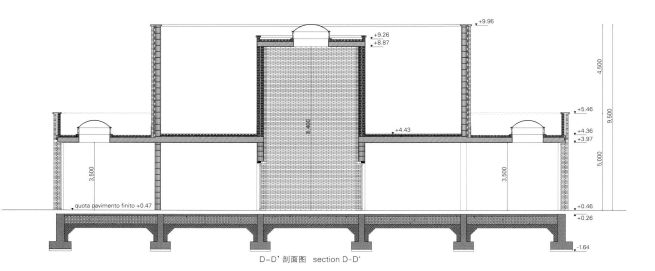
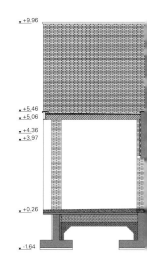

D-D' 剖面图　section D-D'

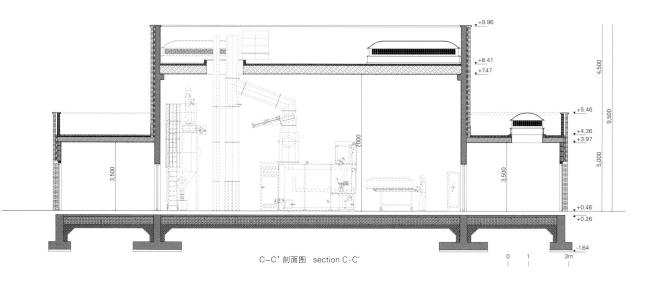
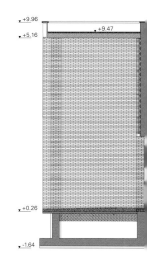

C-C' 剖面图　section C-C'

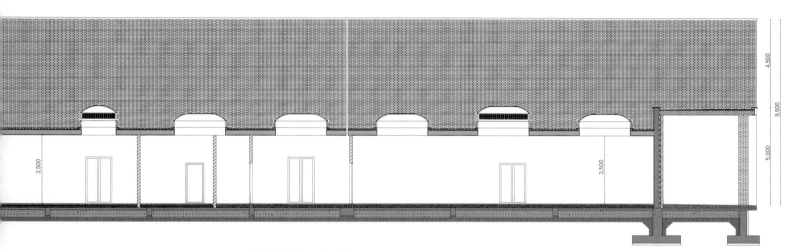

B-B' 剖面图　section B-B'

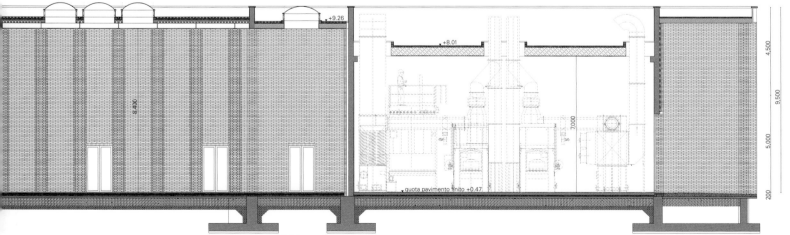

A-A' 剖面图　section A-A'

阿尔塔奇的伊斯兰公墓
Architekt di Bernardo Bader

对于这个项目,建筑师决定把公墓建成一个原始花园,来作为其设计理念,这一主题在宗教领域反复出现。在这样的设计中,公墓场地与周边环境截然分开。为此,他们在草地上嵌入了高度不等的网格状混凝土墙系统。这些混凝土墙构成面向麦加圣地的坟墓的独特空间。封闭空间的概念与同一平面的主题相符。

在靠近入口长长的建筑外墙上,首先迎接访客的是一段装饰了根据伊斯兰传统设计的八角形图案的洞口。建筑内的礼堂空间光影交错,活泼生动。天花板上环形凹槽内置了灯具,给空间增添了额外的光源。

在祈祷室 (mescid) 中,祷告壁龛 (mihrab) 开了一个面向麦加的窗口。三个金属网状的幕帘挂在白色的木质墙壁前面,幕帘上有规律地编织了镀金的木片瓦,拼写出阿拉伯文的"穆罕默德真主"字样。

混凝土墙的颜色使用了黑色和红色颜料;施工时模架留下的洞也被填补。外部混凝土墙表面呈现的表面结构反映出用于模架的锯木板的粗糙纹理 (有3种不同厚度),而内部混凝土墙的表面则十分光滑,体现出内部和外部的细微区别。

公墓的设计基于宗教信仰和丧葬礼仪,而反过来,宗教信仰和丧葬礼仪又传达出许多人们对自然和社会关系的独特理解。为福尔贝格州的穆斯林设计建造的这一新墓地项目也应清楚地表明这一事实。

不管宗教倾向如何,基督教徒和伊斯兰教徒墓地形式上有一个共同点,即公墓是最早的花园。作为真正的"花园",它的特征是土壤的开垦和与地面清晰的界限。第一次修建公园要围合出一块地,并定义与荒野分明的界限。

该项目的设计目的非常开放,总体理念明确清晰。不同高度的墙体嵌板交织在一起,显得十分精致,形成了对墓地和所建结构的围挡。像手指一样的墓穴可以分阶段进行施工,墓地可以一直延伸至原始的自然景观中。

规划的墓地以矮墙作为分界,形成不同的房间。墓地各自被分成较紧凑的区域 (为了使坟墓埋葬有秩有序) 和带坐椅的小房间。所需的一系列设施是从墙体的主题发展而来的。

Islamic Cemetery in Altach

For this project, the architects selected to work with the notion of the cemetery as the primal garden – a recurring theme across religious boundaries – in which a plot of land is clearly set off from its surroundings. To this end they inserted a lattice-like system of concrete walls of varying heights in a sprinkle meadow. These walls create distinct spaces – oriented toward Mecca – for the graves. The concept for the enclosed spaces adheres to the same planar theme.

Next to the entrance in a long outer wall, the visitor is welcomed by an ornamental opening in a wall – bearing an octagonal motif in accordance with Islamic tradition. The congregation space is marked by a lively play of light and shadow. The built-in luminaries positioned in ring-shaped recesses in the ceiling are an additional source of light.

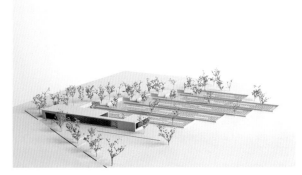

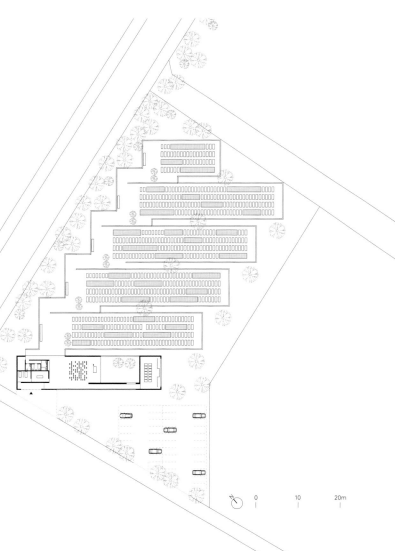

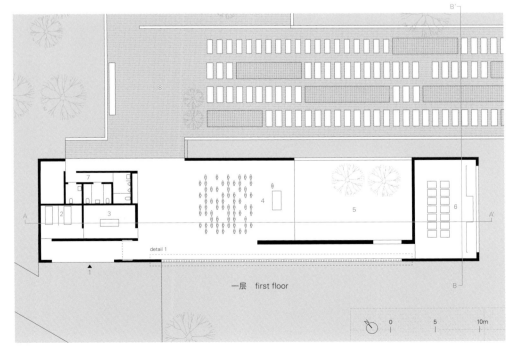

1 入口	1. entrance
2 定线	2. laying-out
3 洗礼区	3. ablution
4 集会区	4. congregation
5 庭院	5. courtyard
6 祈祷室	6. prayer room
7 盥洗室	7. lavatory
8 墓地	8. graveyard

一层 first floor

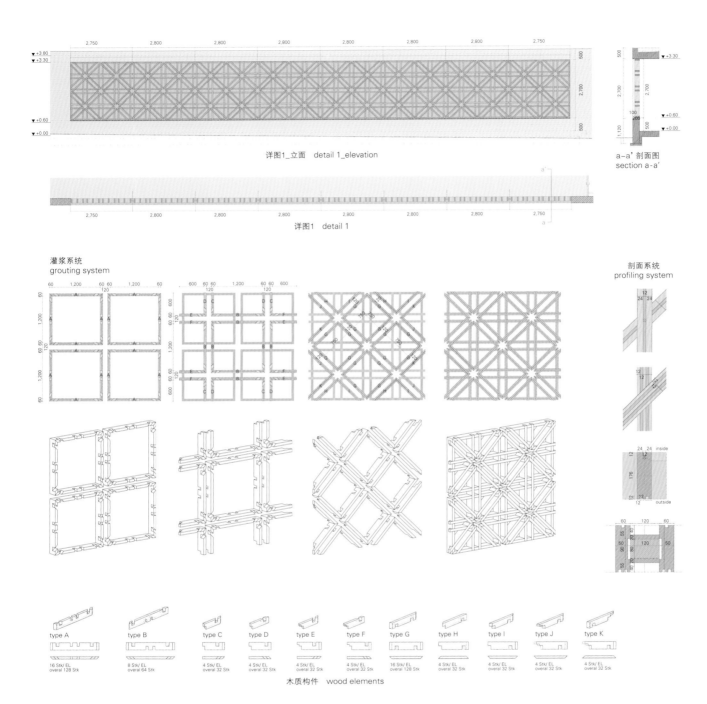

详图1_立面 detail 1_elevation

a-a' 剖面图
section a-a'

详图1 detail 1

灌浆系统
grouting system

剖面系统
profiling system

木质构件 wood elements

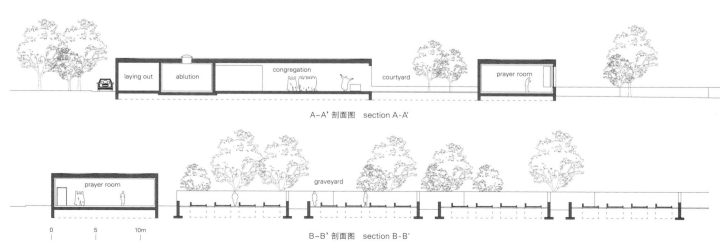

A-A' 剖面图　section A-A'

B-B' 剖面图　section B-B'

In the prayer room (mescid), a prayer niche (mihrab) contains a window facing toward Mecca. Three metal-mesh curtains are situated in front of a white wooden wall, and gilded wooden shingles have been woven into the curtains to spell out the Arabic words *Allah* and *Mohammed*.

The concrete was colored with black and red pigments; the holes left by the formwork ties were filled. The outer concrete surfaces have a surface structure reflecting the texture of the rough-sawn boards (in 3 different thicknesses) used for the formwork; the inner surfaces are smooth, with a subtle distinction between inside and outside.

The design of a cemetery is based on the beliefs and their funeral rites, which in turn say a lot about the particular understanding of nature and social relations. This fact should also be shown clearly in the new cemetery project for Muslims in Vorarlberg.

Regardless of religious orientation, two forms of Christian and Muslim burial site share one point in common, that the cemetery was the first garden. As the real "Urgarten" it is characterized by the cultivation of its soil and its clearly definition from surface. When creating a garden for the first time a piece of land is bounded and clearly delineated against the wilderness.

The aim of the design is a very open and clearly laid-out overall concept. A delicate weave of wall panels in various heights frame the graves and the built-structure. The "finger-like" grave-scale fields allow implementations in stages, and the grave fields extend into the pristine landscape.

The planned grave fields are bordered by low walls and form separate rooms. They are each divided into a compact area for organized grave burials and a small room with sitting-benches. The range of the required facilities are developed from the topic out of the wall.

项目名称：Islamic Cemetery in Altach
地点：Altach, Vorarlberg, Austria
建筑师：Bernardo Bader
艺术家：Azra Aksamija
用地面积：8,415m²
墓地面积：8,400m²
竣工时间：2012
摄影师：courtesy of the architect - p.56~57, p.62, p.62~63
©Adolf Bereuter (courtesy of the architect) - p.59, p.64, p.65
©Peter Allgäuer (courtesy of the architect) - p.61

新希尔维公墓

Giovanni Vaccarini

公墓是一个集体记忆的地方,保存着所经历的重大事件的迹象。走在这些地方,人们会认识到与时间的流逝相比,辛苦的工作只是一个微乎其微的维度。

像在各种生灵所居住的城市中,公墓的扩建也要体现它们所属的那个时代所发生的重大事件。

在近几十年的扩建中,无论是在选址地点的规定中还是个人手工艺品所处的场地中,都很难找出一个设计。这可能反映了在人们大致可以定义为"大规模产业化"时期所经历的不确定性和文化隔阂。

项目所干预的区域位于现在的"圣地"的北面,是以前用于农业耕作(已被遗弃)的一块坡地,现在只保留了一些橄榄树。山谷面朝东面的大海,而西面的主要景观则是人类的杰作——A14高架桥。

这一项目的理念是设计一个与曲线水平进行对话,同时也与周围景观进行对话的墓地体系。在东面和西面,通过使用透明的围栏,人们仍可一览部分海光山色。呈线形排列的建筑面向自然景观,从普遍的一个墓地的建造规则中分离开来。

在意大利文化中,篱笆是定义"圣地"的一个主要元素,它标志着内外的界线,将建筑内部和外部之间的阻碍最小化。该项目旨在支持周围景观中由模架所定义的冥想和回忆空间,是将建筑置于沉寂中的一个灵魂维度。

项目所使用的材料主要有两种:石料和灰泥。几栋建筑物设计为线性体量,可以被看做是一个整体,石质饰面,可以俯瞰周围景观。石质饰面的表面粗糙,颜色不一。外墙面凸凹不平,各处的节点甚至裸露在外,建筑以"绘"有类似于大型条形码的垂直切口为特色。而建筑的内侧(腹部)则采用可直视的混凝土和灰泥材料。

New Silvi Cemetery

The cemetery is a place of collective memory in which the signs of the passing of events are preserved. Walking in these places people will realize that their hard work is a poor dimension compared with the passage of time.

As in the city of living creatures, the operations of expansion of the cemetery reflect the events of the eras they belong to.

It is difficult to find a design in the expansion of recent decades, neither in the settlement rules, nor in the grounds of individual artifacts. It probably reflects the uncertainty and the cultural separa-

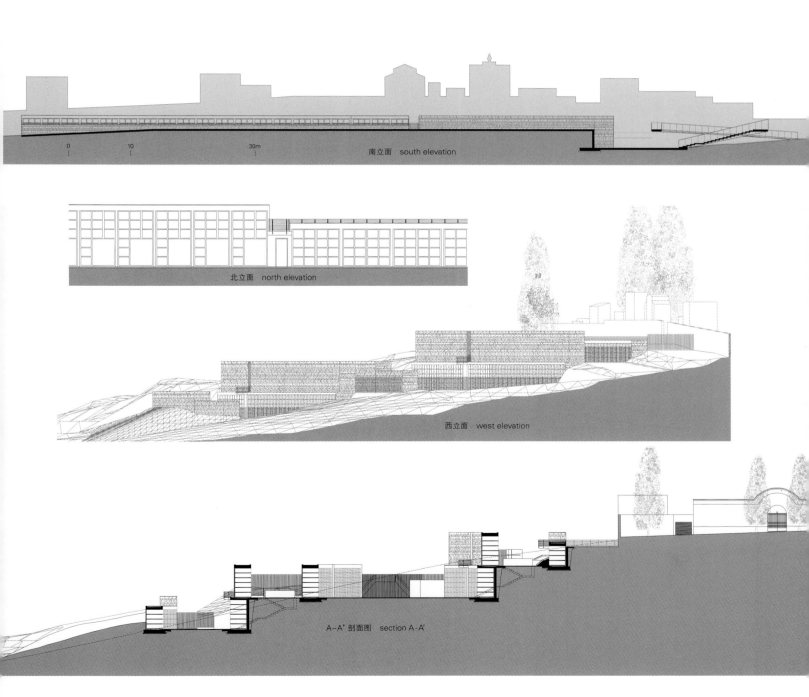

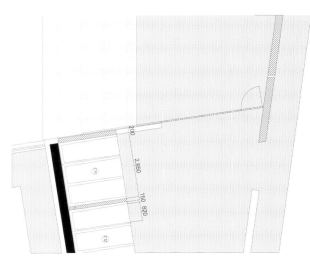

详图1_人行道入口
detail 1 _ pedestrian entrance

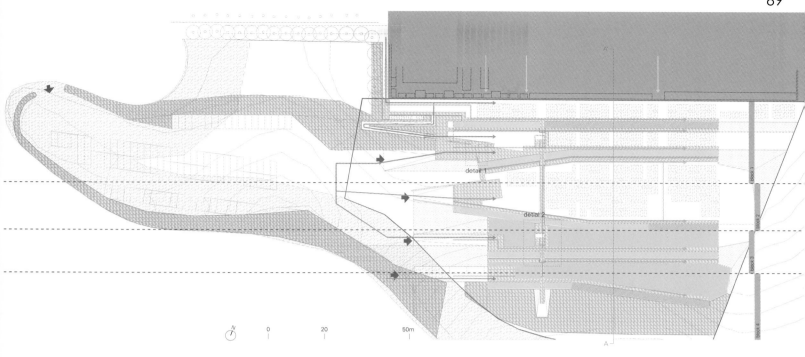

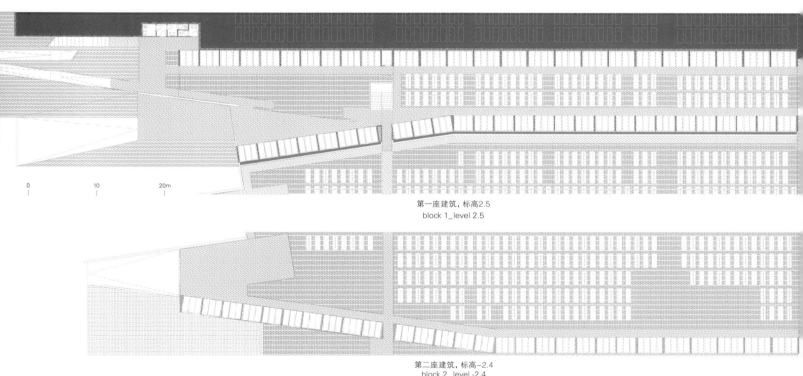

第一座建筑，标高2.5
block 1_level 2.5

第二座建筑，标高-2.4
block 2_level -2.4

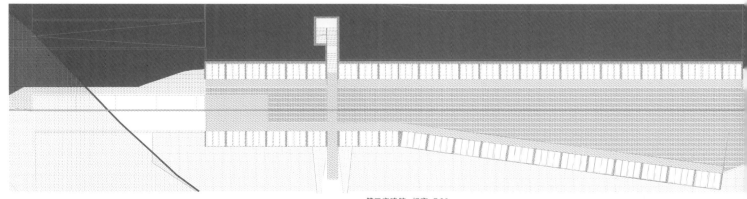

第三座建筑，标高-7.30
block 3, 4_level -7.30

tion that people have lived during what they can roughly define "the mass industrialization".

The area of intervention is located to the north of the present "holy field". This is a slope area abandoned by the agriculture use, in memory of a place where only olive trees remain. The valley opens to the sea to the east; on the west the landscape is dominated by the artifact of the A14 viaduct.

The idea of the project is to design a settlement system that talks with the curves level and at the same time with the surrounding landscape. On the east and on the west, through the transparent fence, they frame portions of the marine and hill landscape. The stringy buildings are opened in the landscape unhinging one of the canons of the cemetery.

The fence is in the Italian culture, one of the main elements of the definition of the "saint field". It marks an inside from an outside, minimizing interference between interior and exterior of the architecture. The project seeks to support the space of meditation and recollection of the frames in the surrounding landscape; it is a dimension of soul in which the building put up the silence.

The materials of the project are essentially two: the coating of stone and plaster. The buildings are designed as monolithic stringy volumes coated in stone, overlooking the landscape. The coating is made with a rough stone with color variations. It is a scraggy use, the joints between the various resorts have been left open; the architecture is marked by vertical cuts that draw a macro bar code. The inside part of buildings (the "belly") is all in face-to-view concrete and gray plaster.

项目名称：New Silvi Cemetery
地点：Silvi, Teramo, Italy
建筑师：Giovanni Vaccarini
承批人和执行人：Consorzio Progetti & Finanza
成员：Cosimino Casterini, Attilio Mauri
结构工程师：Pietro Taraschi
甲方：Local Council of Silvi
用地面积：12,000m² 总建筑面积：5,000m²
有效楼层面积：3,500m²
造价：EUR 2,700,000
设计时间：2007
施工时间：2008~2009
摄影师：©Alessandro Ciampi (courtesy of the architect)

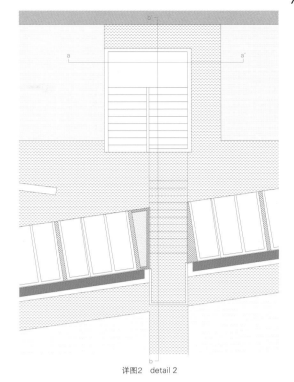

详图2 detail 2

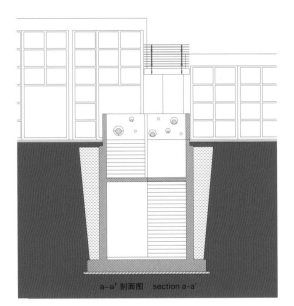

a-a' 剖面图 section a-a'

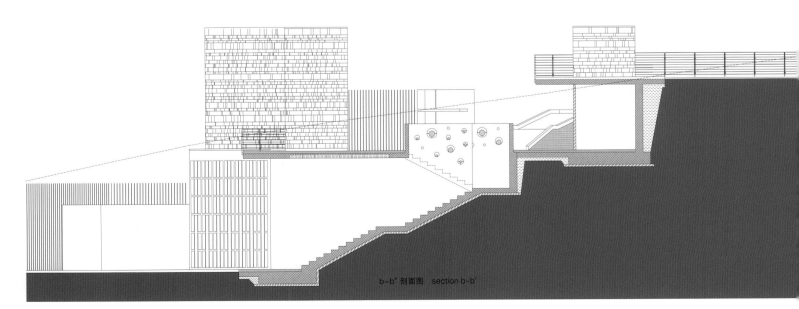

b-b' 剖面图 section b-b'

云神殿
Clavel Arquitectos

这个项目不是一座建筑,也不是一个小结构,甚至不是一个纪念碑。当时的想法是为葬礼定义一个舞台背景,并且考虑了空间和时间情景。

在这个项目中,建筑师注意了两个形象的设计。

外观!

第一个是中世纪时期的无法折叠的门板,过去用来建筑可拆卸的外立面或是祭坛装饰品,现在得到重塑,变成Z字形的曲折立面。死亡让人们想到的是神秘与恐惧。死亡是从一个世界到另一个未知世界的过渡,而这一生死转换过渡的过程是在坟墓中发生的。

因为坟墓是生死的过渡,所以坟墓要永远关闭,它们的门也应该永远不要开启。坟墓无法打开,即使你想打开也无济于事,因此没有门把手或锁。门深深地嵌入墙里,与墙融为一体。因此,坟墓的入口是秘密的,就像里面所发生的生死过渡一样神秘。实际上,门面只有用一种特

场地平面图　situation plan

有的方法才能打开,这种秘密的方法几乎只有主人才知道。

五个梯形门采用着色的平板玻璃饰面来建造,用光滑的不锈钢作为门的框架,门可以围绕隐藏的金属杆旋转。

内部!

内部,建筑师们设计了一团阳光从中掠过的白云:这种大气状况在自然界中才可以看到,与人们所了解的"神圣"文化联系起来,从而刺激人们的心灵。在抽象的白云中,那一刻变得明确而具体,那是道路的尽头,这条道路从地下室开始,向上升至中间平台,一直向上到达云间。人们的视线在此迷失,迷失在充斥这个地方的抽象而虚幻的氛围中。

大片抽象的云由三角形排列的钢板建造,表面涂以白色聚氨酯漆,其白色进一步增强了抽象和光线反射效果。

照明装置

为了获得引人注目的效果,设计中使用了两套照明装置。首先,后面的墙面变得半透明,旨在引入大气照明。这一照明途径确保神殿内最低程度的照明。其次,为了创建一个失重且神秘的环境,建筑顶部设计了三个不规则天窗,自然光线穿过天窗,反弹到金属云层的多个角落中。

天窗和后面的墙面都采用层压玻璃和玛瑙来建造。

地面

最后,神殿里还有一个有趣的地方:地上一层地面与地下一层地面之间的不同,用来象征从一个世界到另一个世界的过渡,从黑暗到光明的过渡。地上一层地面铺的是白色大理石,其颜色和亮度可以加强云的抽象效果;地下一层的地面铺的是与其形成鲜明对比的黑色玄武岩石材。

 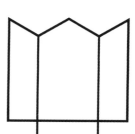 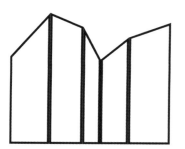 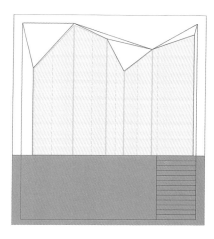

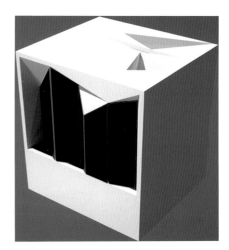
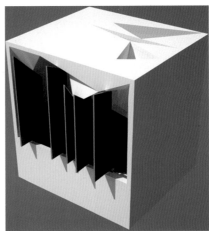

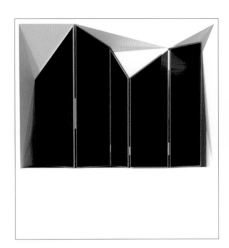
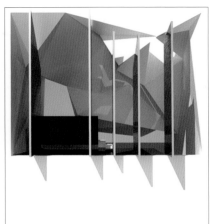

Cloud Pantheon

The project was not for a building, nor a small structure, nor even a monument. The idea was to define a stage setting for a burial, taking into account the spatial and temporal situations.

In this project, the architects worked with two images.

Outside!

The first one was the medieval unfoldable boards, which used to work as removable facades or altarpieces, and now get remade in the zigzag facade. Death inspires human beings with something between mystery and fear. To die means a transition between this world and others that nobody knows. The tomb is where this transition takes place.

According to this situation, tombs should be closed forever, and their doors should never be opened. It should be impossible to open them, even if you want, so there are no handles or locks. The doors insert themselves in the walls, merging together. Therefore, the entrances are secret, just like the transits that occur inside. In fact, the facade can be only opened in a specific way, almost a combination that only the owner knows.

The five trapezoidal doors have been constructed with body-tinted slabbed glass, framed in satin stainless steel, and revolve around a hidden metallic rod.

Inside!

Inside, the architects found a cloud that gets crossed by sunbeams: an atmospheric situation that can be found in the nature and that, linked to our cultural references of the Divine, stimulates our spiritual side. That moment gets crystallized in the abstraction of the white cloud, the end of a way that, starting from the basement ascends to the intermediate platform and goes on until the cloud's space, where people's sight gets lost in the abstract and unreal atmosphere that fills the place.

The big abstract cloud has been constructed by triangulating protected steel plates, finished with white polyurethane paint. Its white color reinforces both the abstraction and the light reflections.

Lighting Mechanisms

Two lighting mechanisms have been used, so as to get eye-catching effect. Firstly, the back wall turns into a translucent surface, aiming to get the entrance of atmospheric lighting. This mechanism ensures a minimum degree of lighting inside the pantheon. Secondly, three irregular skylights on the top let natural light go through and bounce off on the multiple nooks of the metallic cloud, in order to create a weightless and enigmatic ambient.

Both the skylights and the back wall have been constructed with laminated glass and onyx stones.

Pavements

Finally, there is another interesting resource inside the pantheon: the distinction between the pavement above ground level and the one below ground level, as a way to symbolize the transition from one world to another, from darkness to light. White marble is used above ground level, due to its color and brightness, to reinforce the abstraction of the cloud; pavement below ground level is made in contrast with black basalt stone.

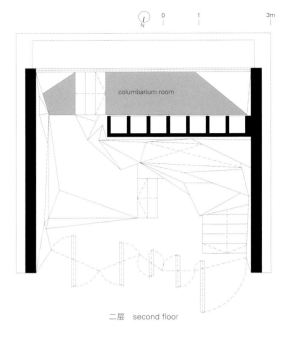

二层 second floor

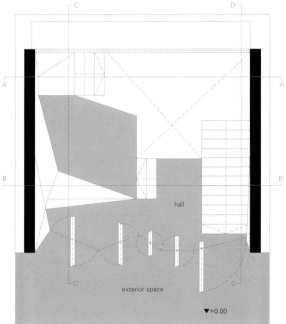

一层 first floor

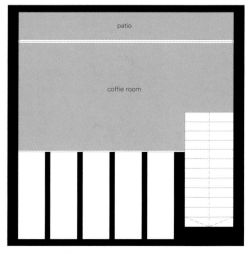

地下一层 first floor below ground

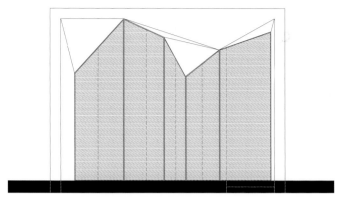
北立面_门关闭 north elevation_door closed

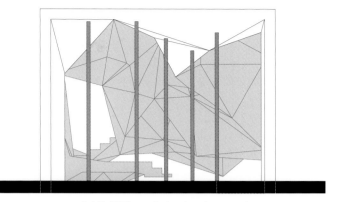
北立面_门开启 north elevation_door opened

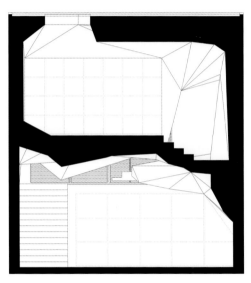
A-A' 剖面图 section A-A'

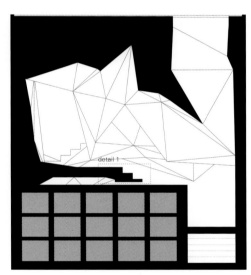
B-B' 剖面图 section B-B'

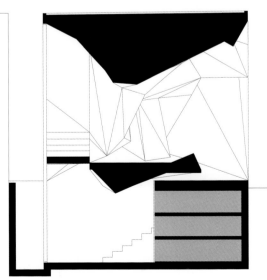
C-C' 剖面图 section C-C'

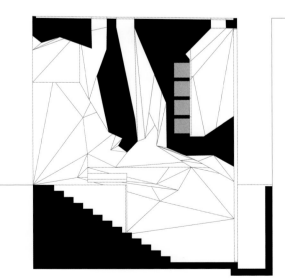
D-D' 剖面图 section D-D'

门详图 door detail

a-a' 剖面图 / section a-a'
b-b' 剖面图 / section b-b'
c-c' 剖面图 / section c-c'
d-d' 剖面图 / section d-d'
e-e' 剖面图 / section e-e'

- 40.40.2 steel tube
- 10mm thick anchor plate
- stiffening gusset
- 150. 100.2 steel tube spatial auxiliary structure (welded connections)
- 10mm thick anchor plate
- 2mm thick folded steel sheet, white lacquered
- steel door shaft, fastened to auxiliary structure

固定钢板门的辅助结构
auxiliary structure for door steel plates fastening

80

项目名称：Cloud Pantheon
地点：Espinardo, Murcia, Spain
建筑师：Manuel Clavel Rojo
合作商：Robin Harloff, Mauricio Méndez Bustos, David Hernández Conesa
结构和机械工程师：Clavel Arquitectos
用地面积：68m²
总建筑面积：38.95m²
有效楼层面积：80m²
竣工时间：2010
摄影师：©David Frutos (courtesy of the architect)

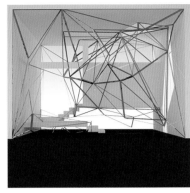

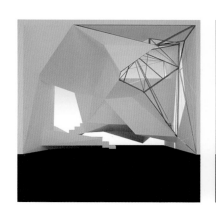

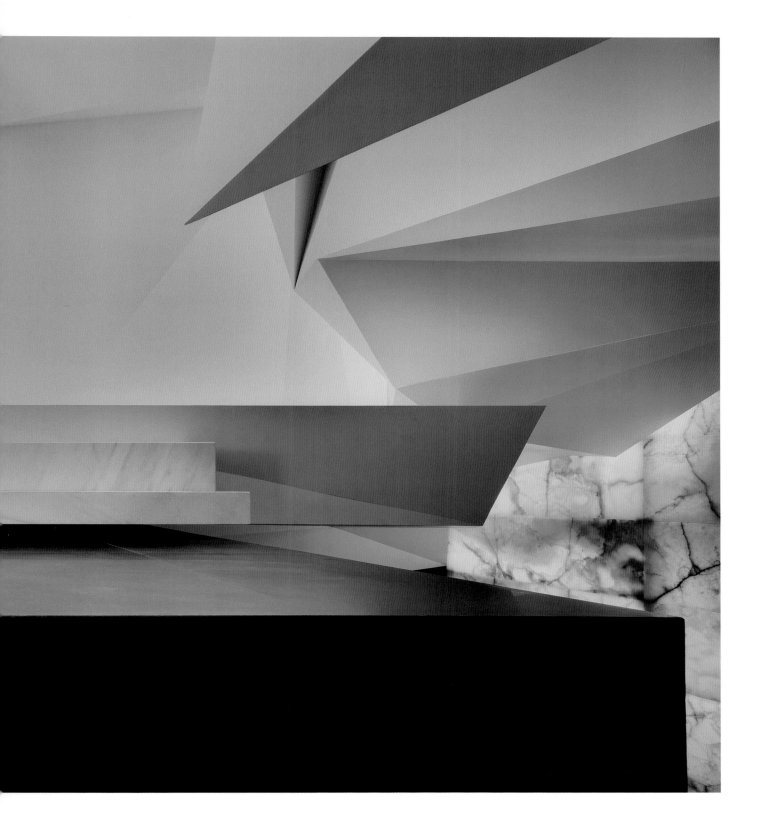

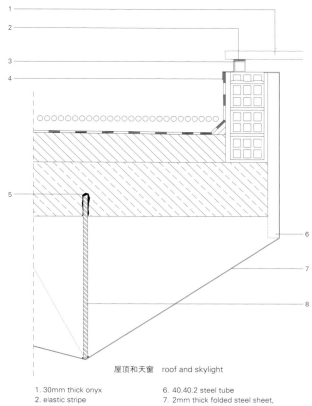

屋顶和天窗 roof and skylight

1. 30mm thick onyx
2. elastic stripe
3. 40.40.3 galvanized steel frame
4. galvanized steel folded sheet
5. epoxy resin for rod anchoring
6. 40.40.2 steel tube
7. 2mm thick folded steel sheet, white lacquered
8. ø16mm corrugated steel anchoring rod

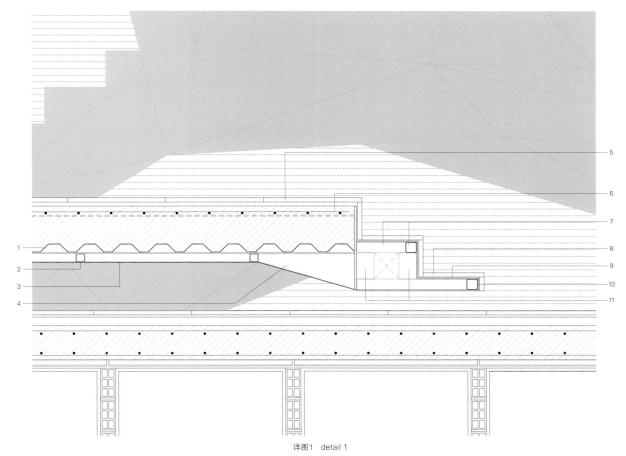

详图1 detail 1

1. composite slab, standing on IPE 180 steel profile
2. 40.40.2 galvanized steel tube
3. 2mm thick folded steel sheet, white lacquered
4. steel gusset, welded to IPE 180 steel profile
5. 20mm thick white macael marble, slip resistant finish
6. reinforcing wire mesh
7. 50.50.3 steel tube auxiliary structure
8. 20mm thick white macael marble, slip resistant finish
9. epoxy resin adhesive
10. 2mm thick steel sheet cladding*
11. 5mm thick stiffening gusset*

* 2-side stainless finish coating

茵格海姆葬礼小教堂
Bayer & Strobel Architekten

茵格海姆市的每个社区都有自己的小公墓,但人们预期仍十分有必要修建新的公墓。计划修建的茵格海姆市新公墓位于市区莱茵河附近的Frei-Weinheim。新公墓和葬礼小教堂的重新设计方案是通过2008年举行的建筑师设计比赛选拔出来的,获奖者是Bayer & Strobel建筑事务所。

安静、祥和、告别

巨大的石块墙是公墓新设计的一个基本元素,由当地典型的黄灰色石头建造。厚重的墙体将公墓与外面的街道隔离开来,形成一个封闭的整体,成为人们沉思默祷的安静所在。同时,这些围墙又起着挡土墙的作用,将公墓的入口区域与稍微抬高的用于安葬的区域分隔开来。

葬礼小教堂是公墓的核心主体。小教堂的外立面和内部都用石料砌成,自然而然地成为入口区域的一部分。公墓内外的一系列空间复杂多样,过渡自然流畅。公墓和葬礼小教堂被认为是一个非常和谐的单元,成为一个不可分割的整体。

殡仪厅不仅是人们哀悼逝者的空间,也是希望之地,本身散发出明亮而友好的气氛。旁边一侧的大型落地窗开向中庭般的庭院,院内景色尽收眼底。为了充分发挥它在整个公墓系统中的重要性,殡仪厅以一个人字形屋顶被清晰地标识出来。这样营造的室内空间,让人感到庄严肃穆,简单而适合,达到了目的。

Ingelheim Funeral Chapel

Each community of the city of Ingelheim has its own small cemetery. Prospective there is a significant extension necessary. The intended new main cemetery of Ingelheim will be at the city quarter Frei-Weinheim near the river of Rhine. A redesign of the new cemetery and funeral-chapel was decided in a competition for architects in 2008, won by Bayer & Strobel Architekten.

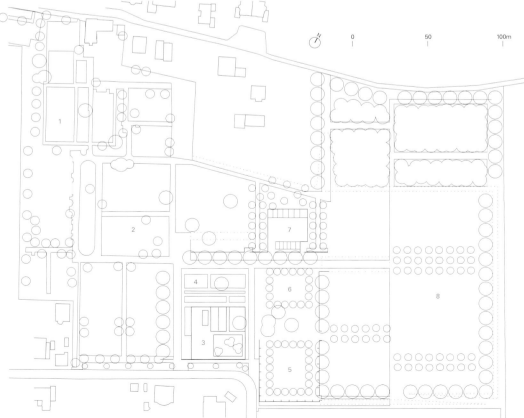

1 历史庭院
2 现存的公墓
3 新建的葬礼小教堂
4 花园
5 南侧广场
6 路边的具有历史意义的十字架
7 维修站
8 未来的公墓场地

1. historic graveyard
2. present cemetery
3. new funeral chapel
4. flower garden
5. southern square
6. historic wayside cross
7. maintenance depot
8. future cemetery

Quietness, Peace, and Farewell

As an essential element of the redesign of the cemetery, massive quarry-stone walls were built, made of the typical local yellow-gray stone. They block off the cemetery from the street, enclosing it as a place of peace and contemplation while also acting as retaining walls separating the entrance area from the more elevated parts of the cemetery used for burials.

As the heart of the cemetery, the funeral chapel is made into a natural part of the entrance area through the use of quarry stone for its facade as well as in the interior. The result is a sophisticated series of interior and exterior spaces with finely modulated transitions. Cemetery and funeral chapel are perceived as a harmonious unit and become an inseparable whole.

Being not only a space of mourning, but also a place of hope, the funeral hall itself exudes a bright and friendly atmosphere, alongside large-scale windows opening the view into the patio-like courtyards. To do justice to its significance within the cemetery complex, it is clearly marked with a gable roof. This creates an interior that feels dignified and solemn as well as simple and appropriate to its purpose.

项目名称：Funeral Chapel Ingelheim
地点：Ingelheim am Rhein, Germany
建筑师：Bayer & Strobel Architekten BDA
结构工程师：Ingenieur-Gesellschaft TRAGWERK Angnes + Rohde mbH
景观建筑师：jbbug Johannes Böttger Büro Urbane Gestalt
甲方：Stadt Ingelheim am Rhein
用地面积：15,700m² (without old cemetery)
有效楼层面积：1,043m²
竣工时间：2012.5
摄影师：©Christian Köhler (courtesy of the architect) - p.84, p.88, p.90^{bottom}, p.92,
©Peter Strobel (courtesy of the architect) - p.86~87, p.90 $^{top-left, top-right}$, p.91

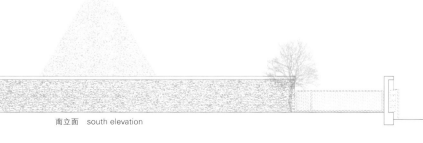

南立面　south elevation

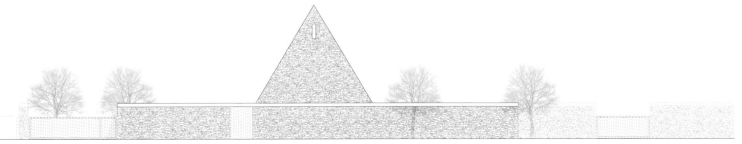

北立面 north elevation

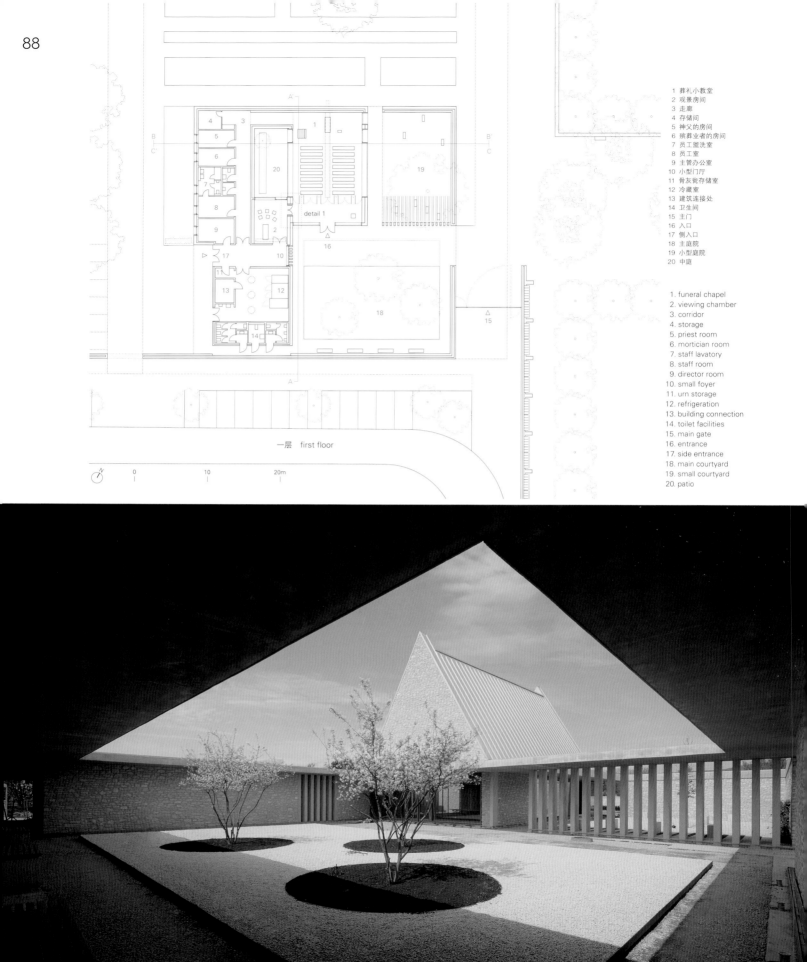

一层 first floor

1	葬礼小教堂	funeral chapel
2	观景房间	viewing chamber
3	走廊	corridor
4	存储间	storage
5	神父的房间	priest room
6	殡葬业者的房间	mortician room
7	员工盥洗室	staff lavatory
8	员工室	staff room
9	主管办公室	director room
10	小型门厅	small foyer
11	骨灰瓮存储室	urn storage
12	冷藏室	refrigeration
13	建筑连接处	building connection
14	卫生间	toilet facilities
15	主门	main gate
16	入口	entrance
17	侧入口	side entrance
18	主庭院	main courtyard
19	小型庭院	small courtyard
20	中庭	patio

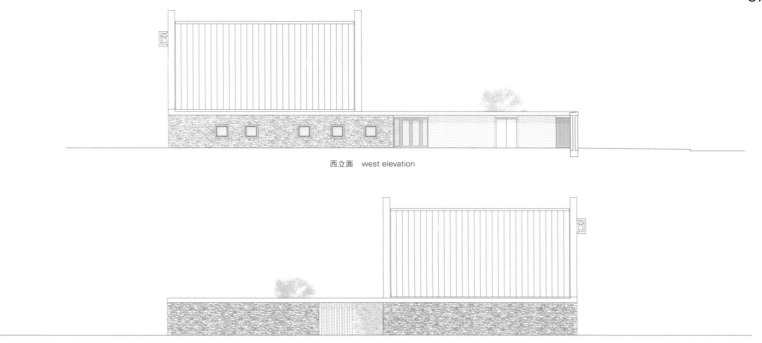

西立面　west elevation

东立面　east elevation

A-A' 剖面图　section A-A'

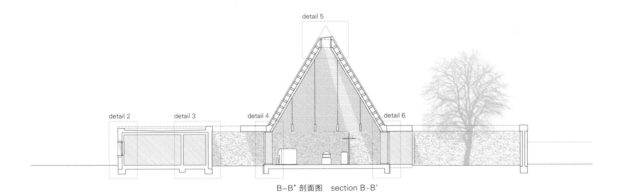

B-B' 剖面图　section B-B'

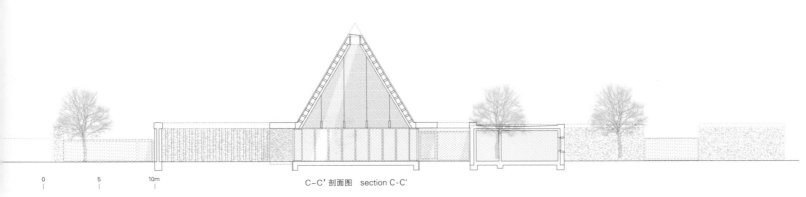

C-C' 剖面图　section C-C'

1. türbänder verdeckt
2. wandvertäfelung:
   mehrschichtparkett eiche, 160-340/15mm
   geschliffen, geölt
   traglattung 40/60mm
   unterkonstruktion metallständerwerk
3. aussparung f. drücker innenl. stoßfänger
4. türlaibung betonfertigteil
5. befestigung/kippsicherung

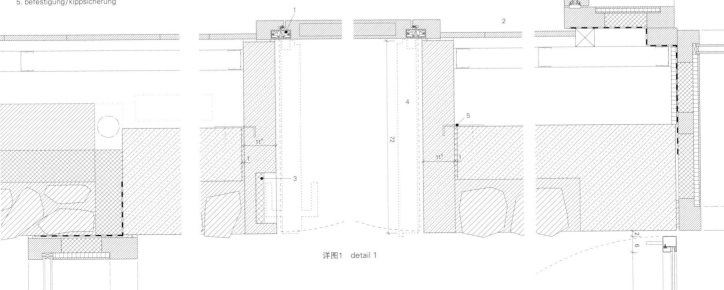

详图1　detail 1

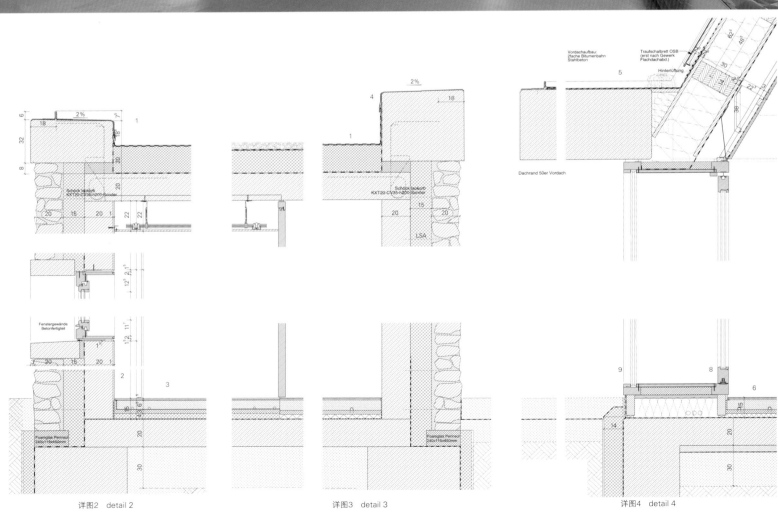

详图2 detail 2　　　　　详图3 detail 3　　　　　详图4 detail 4

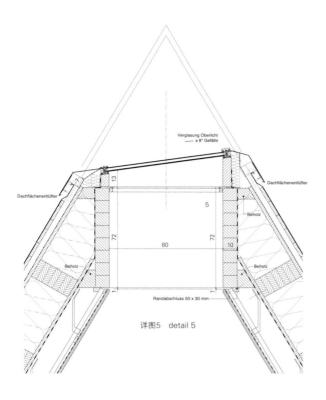

详图5　detail 5

1. flat roof: bitumen sealant layer rough, bitumen sealant layer adhesive, insulation EPS sloped, vapor barrier, reinforced concrete, suspended ceilings
2. wall: plaster, sand-lime brick, insulation EPS, masonry travertin,
3. floor: tiling, floating screed, underfloor heating, system panel insulation, thermal insulation PUR, sealant layer, reinforced concrete
4. coating PU, UV resistant
5. step roof: sheet tinned copper, planking, ventilation layer, OSB panel, roof beam, OSB panel, vapour barrier, suspended ceilings wood
6. terrazzo floor: smoothed terrazzo 90mm, underfloor heating, system panel insulation, thermal insulation PUR, sealant layer, reinforced concrete
7. fairfaced concrete
8. sliding door
9. fixed glazing

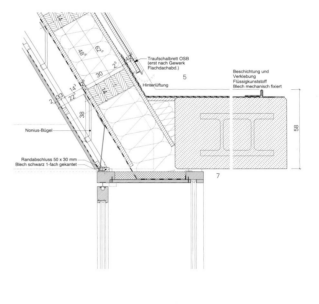

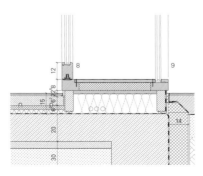

详图6　detail 6

埃伦巴赫公墓访客中心

Andreas Fuhrimann Gabrielle Hächler Architekten

项目名称: Erlenbach Cemetery Building
地点: Erlenbach, Zurich, Switzerland
建筑师: Andreas Fuhrimann Gabrielle Hächler Architekten ETH BSA SIA
项目主管: Regula Zwicky
结构工程师: Reto Bonomo
电气工程师: Kowner AG
甲方: Germeinde Erlenbach
用地面积: 8,346m²
有效楼层面积: 200m²
规划时间: 2008~2009
施工时间: 2009~2010
摄影师: ©Valentin Jeck (courtesy of the architect)

从结构上来说，新公墓访客中心的位置既包含了位于苏黎世湖边上的公墓场地，也包含了与此平行并置的教堂，成为这两栋建筑沟通对话的渠道，横跨两者之间的墓地。与教堂截然相反，集多功能为一体的公墓访客中心的设计所烘托的是公墓安静、私密的氛围。访客中心的所有功能都集中在一个巨大的屋顶下，无拘无束。面向公墓一侧的走廊镶有玻璃，人们通过镶有玻璃的走廊可以到达坐落在建筑中间的殡仪馆，殡仪馆设有几个房间。绿色且大多是不透明的玻璃形成一道视觉屏障，成为人们即将要面对逝者的一个重要的过渡区。胡桃木墙面板和天窗的自然照明使殡仪馆庄严而神圣，同时让人感到温暖而舒适。空间的内向性满足了人们沉默哀悼、与死者告别的需要。

访客接待室和带有屋顶的外部空间面向俯瞰景色优美的苏黎世湖以及举行一些小仪式的区域。两侧带有穿孔图案的装饰混凝土墙营造了一种具有保护性但同时庄严肃穆的气氛。这一古色古香的混凝土结构与玻璃立面的轻盈形成明快的对比，具有抽象的象征意义。玻璃立面的颜色或是绿色，或是棕色，深浅不一，使建筑成为自然环境的一部分，也体现了不同文化背景中神圣的宗教建筑使用彩色玻璃的悠久传统。新公墓访客中心的设计旨在帮助哀悼者正视并接受他们所处的特殊情况，为所有忏悔的默哀仪式提供一个享有尊严的现代环境。

Erlenbach Cemetery Building

The situation of the new cemetery building constructionally brackets the grounds of the cemetery directly at the Lake of Zurich and the parallel lie of the church, placing the two buildings in a dialogue with each other and spanning the graveyards between them. As opposed to the church, the functionally hybrid building is conceived as a pavilion to serve the secluded, intimate atmosphere of a cemetery. All of the uses are combined in a free form under a massive roof. Situated in the center are the rooms of the funeral parlor, which are accessed via a glazed corridor facing the cemetery. The green, mostly opaque panes provide a visual

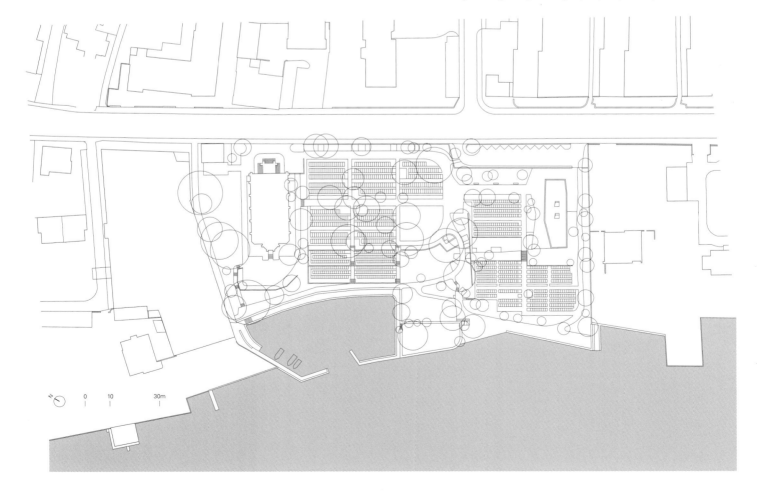

barrier, and the space forms an important interim area in which to prepare for the encounter with the deceased. The walnut paneling and the natural zenithal lighting give the funeral parlor an appropriate dignity, but also a warmth and comfort. The introversion of the spaces provides the silence for mourning and taking leave of the deceased.

The visitors' room and the covered exterior space are orientated towards the most beautiful point overlooking the lake, and where small ceremonies can be held. Two perforated, ornamental concrete wall elements in the covered outside area create a protective but simultaneously solemn atmosphere. The archaic concrete construction contrasts pleasurably with the lightness of the glass facade, creating an abstract symbolism. The coloring of the panes of the facade in various tones of green and brown makes the pavilion part of the natural surroundings and is tied to a long tradition of applying colored glass in the sacred buildings of different cultures. The architectural expression of the building is intended to help the mourners in coming to terms with the exceptional circumstances in which they find themselves, and to provide a dignified and contemporary setting for the mourning rituals of all confessions.

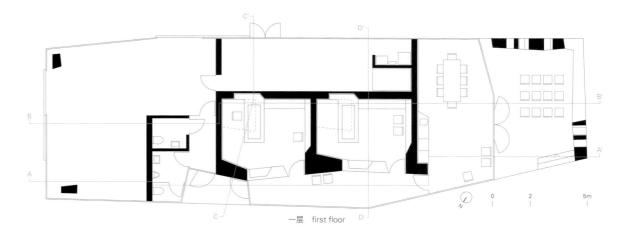
一层 first floor

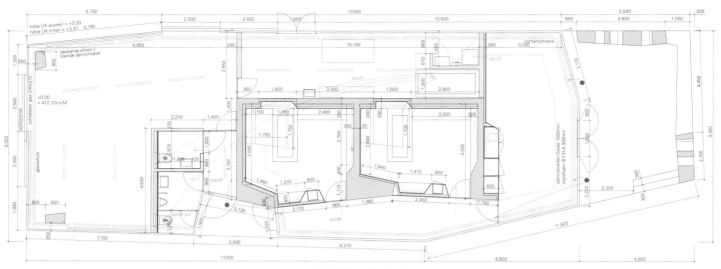

天花板详图　ceiling detail

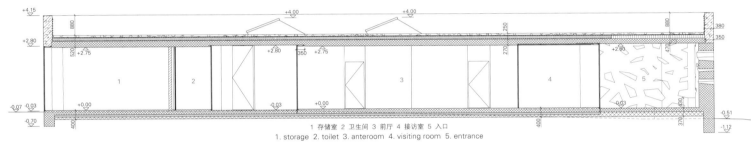

1 存储室 2 卫生间 3 前厅 4 接访室 5 入口
1. storage 2. toilet 3. anteroom 4. visiting room 5. entrance

A-A' 剖面图   section A-A'

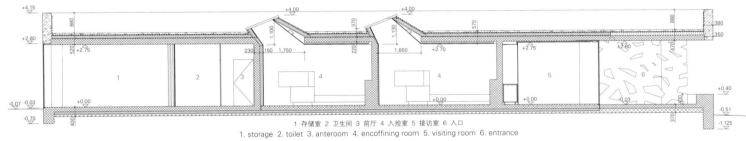

1 存储室 2 卫生间 3 前厅 4 入殓室 5 接访室 6 入口
1. storage 2. toilet 3. anteroom 4. encoffining room 5. visiting room 6. entrance

B-B' 剖面图   section B-B'

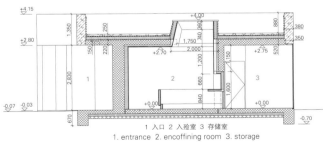

1 入口 2 入殓室 3 存储室
1. entrance 2. encoffining room 3. storage

C-C' 剖面图   section C-C'

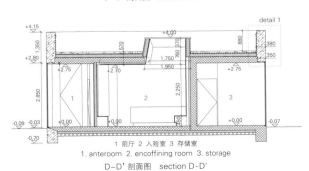

1 前厅 2 入殓室 3 存储室
1. anteroom 2. encoffining room 3. storage

D-D' 剖面图   section D-D'

Abschluss Brüstung (Liapor-Dämmbeton):
−2% Gefälle (ohne Blech!)
Anschluss Spengler-Detail/OK Brüstung:
Flüssigkunststoffabdichtung oder Alternativ
Schutzanstrich Kautschuk- oder 2-Komponentenanstrich
Aufbordung innen:
Beschieferte Elastomer-Bitumen-Dichtungsbahn

Flüssigkunststoff oder Schutzanstrich

flachdachaufbau:
extensivsubstrat 80mm
bitumendichtungsbahn 2lagig: PBD EP 5 wurzelfest / PBD EGV 3
evtl. trennlage
wärmedämmung 140mm: swisspor PUR alukaschiert
dampfsperre: PBD EVA 35
betondecke 270mm

notüberlauf (aussparung 9.5 x 50 cm, mauerstärke 38 cm):
höhe: UK überlauf = +3.37 innen / +3.33 innen
anzahl: 2 stück an SO-fassade, lage siehe grundriss

详图1 detail 1

Kedainiai第一火葬场

G.Natkevičius & Partners

这是在立陶宛建造的第一个火葬场。建立火葬场的想法可追溯至两次世界大战期间,但当时,考纳斯医院只修建了熔炉,用来减少医院的医疗废物。随着火葬传统日盛,立陶宛人在拉脱维亚和波兰都开始使用火葬服务。

在立陶宛建造第一个火葬场并非易事。尽管热情很高,但是直到2011年人们才克服了天主教思想、政治方面的伪善行径以及环境及其他法律的缺乏和不完善带来的种种不便。Kedainiai火葬场所有人、医生和环保主义者终于显示了他们的实力。

建筑地点位于立陶宛中部的工业城镇Kedainiai,这里有31 000名居民。建筑选址位于工业园区。新建筑周围都是杂乱无章、毫无美感的工业建筑:烟囱中浓烟滚滚的制糖厂和肥料工厂。所以建筑的环境甚至看不到一丝神圣感。

在设计过程伊始,建筑师就分析了其他国家的经验。这些具有启发性的实例包括柏林的特雷普托火葬场、德累斯顿火葬场以及安藤忠雄在日本岐阜县设计的火葬场。这些实例都让人印象深刻,这些纪念性建筑宏伟而庞大,拥有振奋人心的场景。然而,Kedainiai的火葬场却是一个占地仅770平方米的小型建筑,要想创造神圣感都让人几乎无从下手。

不甚美观的工业环境引发了建筑师去设计一处简洁甚至有苦行意味的场所。

这是一座单层混凝土建筑,外部与内部均采用了混凝土表面,创造出统一之感。为了与周围的工业环境保持距离,建筑被设计为封闭式的,如同一个性格内向的人。甚至连能引起人们糟糕感受的烟囱都隐藏在建筑的体量之内。设计的主要目标是创建日式的内部庭院,入口处前方栽种了一棵随风摇曳的小榆树。院子创建了一处私密的空间,以及进入该建筑的聚集区域。

这一建筑如同情感过滤器,缓解人们的紧张情绪与压力。

火葬场的室内功能区如下:庭院、休息大厅、两个最终处理大厅、火葬大厅、带烟囱的火化设备房。室内表面由四种材料建造:混凝土、胶合板、铝框玻璃与白色灰泥。苦行风格的内部能让逝者的家人专注于悲伤肃穆的情绪,不会受到任何色彩或细节设计的干扰。每一个人及其脸庞都会成为室内非常重要的一部分。

三种等级的现代火化和空气净化设备由著名德国公司IFZW生产,均符合最严格的环境要求。建筑内部设备齐全,可实施双线火化。

The First Crematorium in Kedainiai

This is the first crematorium in Lithuania. The idea to build a crematorium was born in the interwar period, but at that time only a furnace to reduce medical waste in Kaunas hospital was built. As for the increasing cremation traditions, Lithuanians use cremation services in Latvia and Poland.

To make a path for the first crematorium in Lithuania wasn't easy. Despite of the big enthusiasm, the catholic mind and political hypocrisy, the lack and imperfections of environmental and other

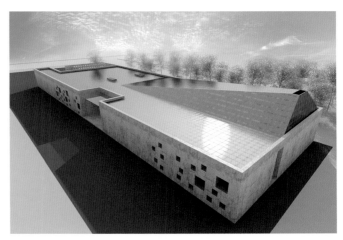

项目名称：The First Crematorium in Kedainiai
地点：Metalistų st., Kèdainiai, Lithuania
建筑师：G.Natkevičius, A.Rimšelis
结构工程师：JSC Constr, Adomas Sabaliauskas, JSC Konstruktoriu Cechas, Kestutis Matijosaitis
用地面积：7,550m²
总建筑面积：775m²
有效楼层面积：789.31m²
竣工时间：2011
摄影师：©G.Česonis (courtesy of the architect)

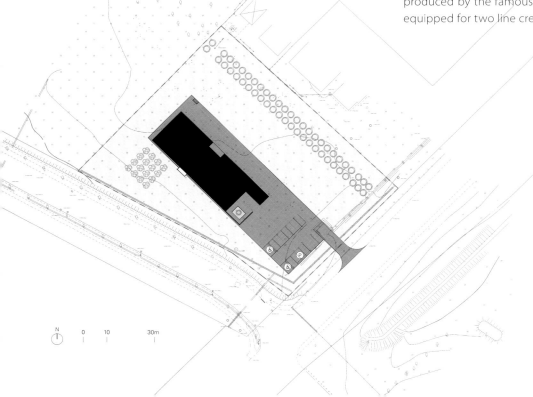

laws were overcome only in 2011. The owners of Kedainiai crematorium, doctors and environmentalists, showed the strength.

The building's site is in the industrial town called Kedainiai, with 31,000 inhabitants in the center of Lithuania. The area for the building was chosen in industrial park. The new building is surrounded by a chaotic, unaesthetic industry: sugar mills, fertilizer factories with smoky chimneys. So the surroundings don't even have a smell of sacred place.

At the beginning of the design process the architect analyzed the experiences of other countries: inspiring examples of crematoriums such as Treptow Crematorium in Berlin, Dresden Crematorium and the crematorium designed by Toyo Ito in Japan, Gifu. These examples were impressive, large monumental buildings with inspiring scenario, however the crematorium in Kedainiai was a building of 770m² and there was almost no place to create sacred script.

Unaesthetic industrial environment provoked to create minimalistic and even ascetic scenario.

It is one story concrete building of which external and internal quality and unity were created with concrete surfaces. In order to distance itself from the industrial environment the building was designed closed like a human introvert. Even the chimney, which causes bad feelings, is hidden in the volume of the building. The main goal of the script is to create the inner Japanese style courtyard with a growing pendulum elm before the main entrance. The yard creates an intimate space, and the accumulation zone before entering the building.

It creates an emotional filter to reduce human experience of stress. The crematorium's interior scenario consists of a courtyard, a lobby with resting area, two final disposition cremation halls and a cremation equipment room with chimney. The interior is created with four surfaces: concrete, wood veneer, glass with aluminum and white plaster. The ascetic inside allows families to concentrate on a solemnly sad hour with no interference of colors and details. Every man and his face become very important parts of the interior.

Three levels of modern cremation and air cleaning equipment, conforming to the strictest environment requirements, were produced by the famous German firm IFZW. The building is fully equipped for two line cremation.

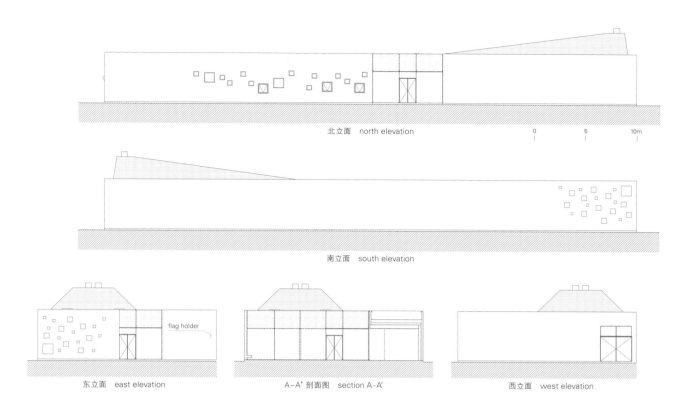

北立面 north elevation

南立面 south elevation

东立面 east elevation   A-A' 剖面图 section A-A'   西立面 west elevation

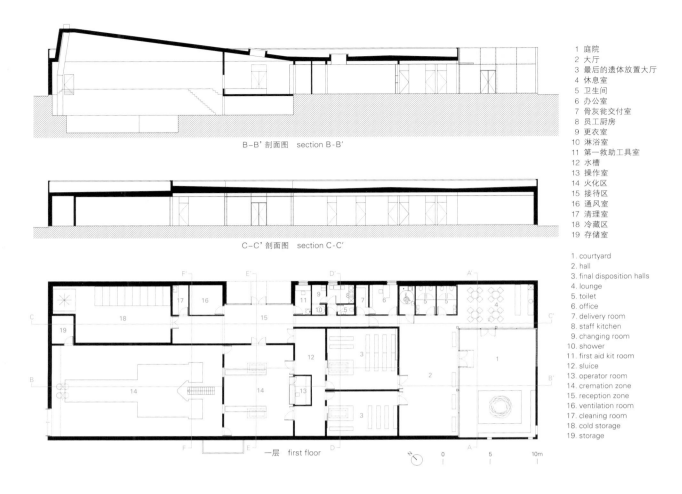

B-B' 剖面图 section B-B'

C-C' 剖面图 section C-C'

一层 first floor

1 庭院	1. courtyard
2 大厅	2. hall
3 最后的遗体放置大厅	3. final disposition halls
4 休息室	4. lounge
5 卫生间	5. toilet
6 办公室	6. office
7 骨灰瓮交付室	7. delivery room
8 员工厨房	8. staff kitchen
9 更衣室	9. changing room
10 淋浴室	10. shower
11 第一救助工具室	11. first aid kit room
12 水槽	12. sluice
13 操作室	13. operator room
14 火化区	14. cremation zone
15 接待区	15. reception zone
16 通风室	16. ventilation room
17 清理室	17. cleaning room
18 冷藏区	18. cold storage
19 存储室	19. storage

D-D' 剖面图　section D-D'

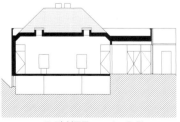
E-E' 剖面图　section E-E'

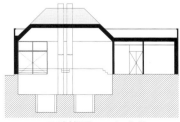
F-F' 剖面图　section F-F'

皮诺索殡仪馆和花园

Cor & Asociados

苏格拉底说:"对死亡的恐惧是出于我们认为自己很聪明而实际上并非如此,假装了解我们对此一无所知的东西。死亡可能是人类最伟大的祝福,没有人知道,但每个人都害怕,好像我们知道并绝对肯定死亡是所有令人不愉快的事情中最糟糕的一件。"

从历史上看,人们发现了对死亡的不同定义。这些不同定义表明人们对死亡的看法发生了转变。从人们认为死亡就是黑暗、痛苦和恐惧,到现在人们把死亡与悲伤、变化和光亮联系到一起。

死亡是人的生存过程中最未知的阶段,设计一幢人死后可能在其中生活居住的建筑必然会把不确定性这一大前提作为设计构思过程中的设计理念。

人们把这座建筑解读为前来悼念的人永远也不会忘记的地方,会永远萦绕于他们脑海的地方,因此设计中要很好地控制敏感领域,充分考虑声音、温度、光线、湿度、照明、隐私、与自然环境的关系等参数,这些很重要。

项目地块位于城镇郊外,一条小巷的尽头,靠近市体育中心,前面是文化中心。体育中心和文化中心都是人们经常活动的地方,而殡仪馆与人们进行的体育活动和文化活动毫不相干,如果将殡仪馆建于这些建筑之中,就会让人感到紧张不安。在这种情况下,建筑师提出整体设计方案,用大量植被绿地突显殡仪馆,使其成为所有这些公共建筑和活动的"中心"。他们设计了一个由29棵日本枫树构成的树林,能够很好地衔接、区分和限定每个部分的不同用途,使这一新建筑成为所有其他活动的组成部分。

另外,殡仪馆的背面被埋起来,如同一个洞穴,它的主立面从地面隐现出来。屋顶的植物绿化与周围环境完全一样,更加强调了这一效果,并能使每天使用邻近体育中心进行活动的学龄儿童看不到殡仪馆。也正是因为这个原因,整座建筑物围绕6处洼地(可被视为庭院)而建,它们能够控制与外部的互动关系。人们在室内房间所能看到的只有天空和房间内部的一切。这种内外的渗透性在这个城镇的新场地中变得非常重要。

这是一个建于正遭受危机的国家的公共建筑。建筑师不应该忘记为修建这栋建筑所付出的努力:殡仪馆占地面积为495m²,但预算只有431 583欧元,这种情况迫使建筑师寻找物美价廉的材料解决方案、建筑技术和维护保养系统,使预算不那么捉襟见肘。同时从周围环境的角度来说,还要考虑最大程度的生态适应性和可持续性。这是一项"投入少但获得多"的干预工作。

Pinoso Funeral Home and Garden

Socrates said *"The fear of death is about considering ourselves wise without being it, since it is pretending to know about what we don't know. Death could be the greatest blessing of human beings, no one knows, and yet everyone fears as if we knew with absolute certainty that it's the worst of evils."*

Historically People find different definitions of death that demonstrate how this concept has moved from positions closer to darkness, pain and fear, to positions related to the concept of sadness, change and light.

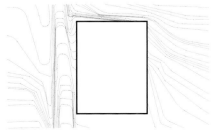

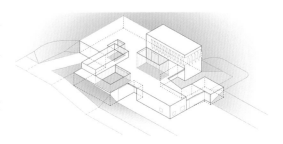

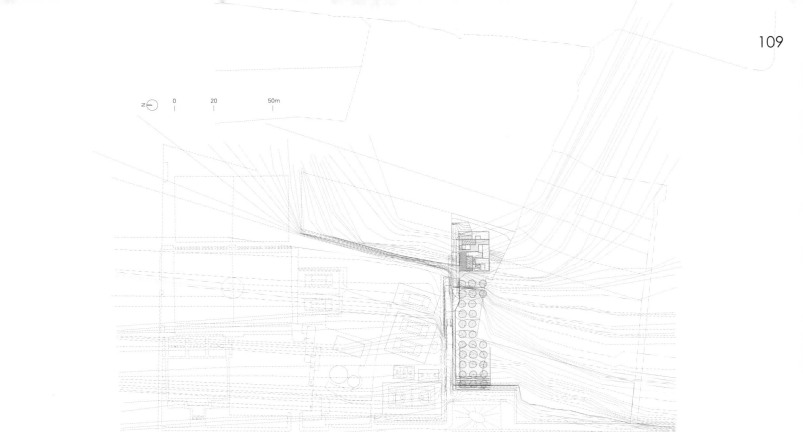

北立面 north elevation

西立面 west elevation

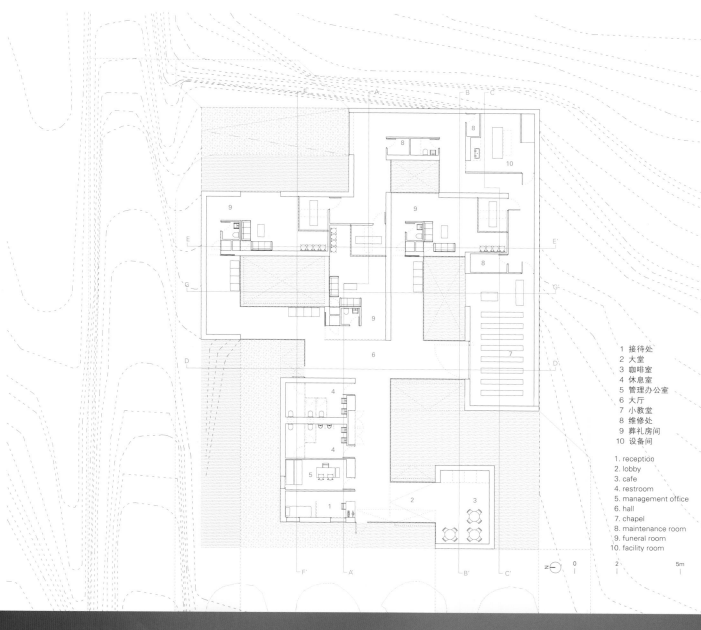

1 接待处	1. reception
2 大堂	2. lobby
3 咖啡室	3. cafe
4 休息室	4. restroom
5 管理办公室	5. management office
6 大厅	6. hall
7 小教堂	7. chapel
8 维修处	8. maintenance room
9 葬礼房间	9. funeral room
10 设备间	10. facility room

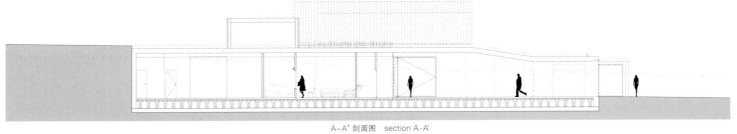
A-A' 剖面图　section A-A'

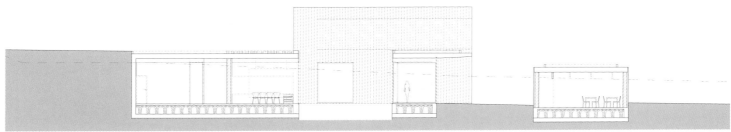
B-B' 剖面图　section B-B'

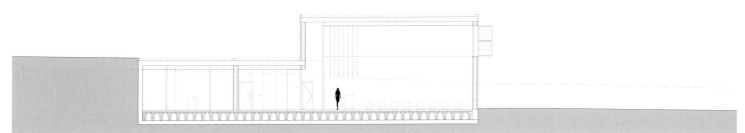
C-C' 剖面图　section C-C'

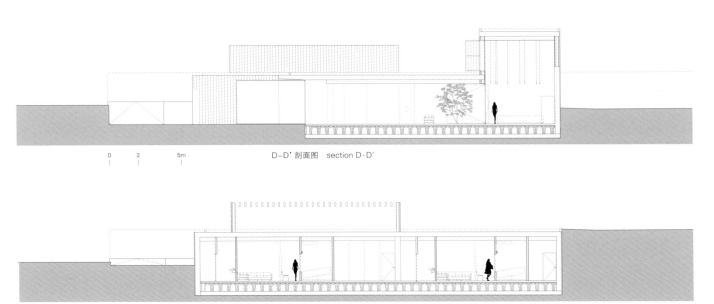

D-D' 剖面图　section D-D'

E-E' 剖面图　section E-E'

项目名称：Funeral home and garden in Pinoso
地点：Pinoso, Alicante, Spain
建筑师：Miguel Rodenas, Jesús Olivares
合作商：Oscar Carpio Rodríguez (Constraction's Project Phase),
José M. Noguera Pardo (3D views)
建筑工程师：José Verdú Montesinos
甲方：Town Council of Pinoso, Alicante
用途：public building, funeral home
用地面积：11,425m²
面积：495m²
造价：EUR 431,583
竣工时间：2010
摄影师：©David Frutos (courtesy of the architect)

Designing a building where people will live, perhaps, the most unknown stage of human existence necessarily involves the assumption of uncertainty as a concept to include in the process of ideation.

People understand this building as a place that will resist being forgotten, remaining in the retinas of their users, and therefore a place where the sensitive realm has to be controlled. Parameters such as sound, temperature, light, humidity, lighting, privacy, relationship with nature become very important.

The plot is situated on the outskirts of town, at the end of a cul-de-sac, closed to the municipal sports center and behind a cultural center, both with great activity. This creates certain tension, since the building is in the middle of various incompatible activities. In this situation the architects proposed to arrange the ensemble generating a vegetation mattress with enough identity to establish itself as the "center" of all these public buildings and activities. They have created a forest of 29 Japanese maples, able to articulate, differentiate and limit the variety of uses. It's a forest where the new building can be a part of all the other activities.

Additionally, the back of the building is buried and, as if it was a cave, its main facade emerges from the fields forward. The vegetal roofs composed exactly the same as its surroundings emphasize this and prevent glances from the adjacent sports facilities used daily by school children for several activities. It is for this reason that the building is arranged around six hollows formed as courtyards, they allow a controlled relationship with the exterior interaction. From the interior rooms people are only able to see the sky and the inside. And this interior and exterior permeability becomes very important in this new town site.

It's a public building in a crisis country. The architects should not forget the effort that this building means for its citizens: the project has 495 square meters and a budget of 431,583 euro, this situation forced them to find material solutions and building techniques, systems of maintenance that reduce that effort. Not forgetting, maximum degree of ecological adaptation and sustainability at the landscape level. This is an intervention that gives more for less.

Cor & Asociados

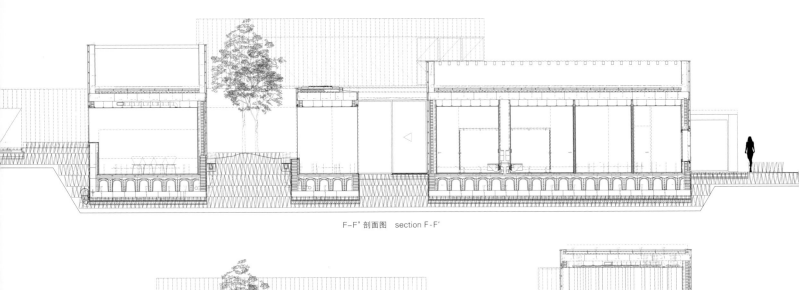

F-F'剖面图　section F-F'

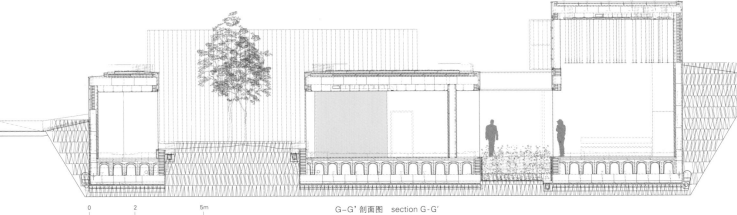

G-G'剖面图　section G-G'

纪念性建筑
设计某种缺失的元素
Architecture of memorial
Designing the Presence of Something Absent

1910年，阿道夫·洛斯发表的言论引起了争议，即只有一小部分建筑属于艺术：坟墓和纪念碑。其余的建筑都只是满足一项功能，不属于艺术领域。洛斯的言论产生了一个严肃的问题，即对艺术和建筑的社会关联性所进行的评估。然而，有一个问题大家则达成了共识：为一座建筑所付出的努力比设计一座坟墓或者纪念碑，或者，比设计任何一个纪念物都更具有挑战性。建筑因此成为一个回应麻烦问题的媒介。那么，什么才能真正地代表过去呢？我们能，或者应该对过去怀抱忠诚的态度吗？我们怎样利用常规的纪念物来描绘非凡的事物？

设计一座纪念碑可以推动集体记忆的发展。纪念碑与历史学家的角色具有相同的目标。然而，如同法国哲学家保罗·利科所说，历史是真相，而记忆则是对应该缅怀的、应该遗忘的以及应该原谅的事件的忠诚体现。因此，根据科利所说，我们必须考虑到纪念碑的功能首先是要提高人们对历史的意识。在记忆、忘记以及原谅之间做出正确的平衡，是一名建筑师或者艺术家在设计纪念物时必须面对的主要挑战。洛斯的不言而喻的言论（即解除一座纪念碑的功能性价值）似乎不承认建筑的愈合潜能。展现某些缺失的事物，这一极大的益处似乎无法由其言论来推动。

In 1910, Adolf Loos has provocatively written that *"only a very small part of architecture belongs to art: the tomb and the monument. Everything else that fulfills a function is to be excluded from the domain of art."* Loos' blunt statement brings about critical issues on the assessment of the social relevance of both art and architecture. However, one thing most would agree: There is arguably a more challenging architectural endeavor than that of designing a tomb or a monument, or, at any rate, designing any kind of memorial. Architecture then becomes the medium that has to answer problematic questions. What is the right representation of the past? Can, or should we be faithful to the past? How can we portray something extraordinary using ordinary tokens?

Designing a memorial is being responsible for contributing to the development of collective memory. It shares, thus, a common goal with the role of the historian. However, as the French philosopher Paul Ricoeur argued, history is about "truth" whereas memory is about "faithfulness" to what *ought* to be remembered, what *can* be forgotten, what *might* be forgiven. Hence, following Ricoeur, we have to take into account that the function of a memorial is, above all, to raise historical consciousness. The right balance between remembering, forgetting and forgiving is thus the main challenge that an architect/artist has to face when designing a memorial. Loos' tacit dismissal of the functional value of a memorial seems then to fall short acknowledging architecture's healing potential. It fails to pay tribute to the immense usefulness of bringing about the presence of something absent.

废奴纪念馆_Memorial to the Abolition of Slavery/Wodiczko+Bonder
911国家纪念馆_National September 11 Memorial/Handel Architects
墨西哥暴乱遇难者纪念馆_Memorial to Victims of Violence in Mexico/Gaeta-Springall Arquitectos
集中营纪念馆_Kazerne Dossin/awg Architecten
Zanis Lipke纪念馆_Zanis Lipke Memorial/Zaigas Gailes Birojs
帕尔米拉纪念博物馆_Memorial Museum in Palmiry/WXCA

纪念性建筑：设计某种缺失的元素_Architecture of Memorial: Designing the Presence of Something Absent/Nelson Mota

欧洲犹太人大屠杀遇难者纪念碑，德国柏林，Peter Eisenman，2004年
Memorial to the Murdered Jews of Europe, Berlin, Germany by Peter Eisenman, 2004

个人经历和集体记忆

大屠杀可以说是最悲惨的悲剧，它一直弥漫在同时期的集体记忆中。期间为此建立了无数个纪念物，本期在前文中也提及了若干个。而最应该被探讨的则是Peter Eisenman设计的欧洲犹太人大屠杀遇难者纪念碑，于2005年在柏林落成。正如Esther da Costa Meyer深刻地指出，Eisenman的设计重新定义了纪念碑的作用，将其作为一个能够唤醒而非代表一个与个人经历不相匹配的事件的媒介。纪念物无法成为可以治疗伤口的宣泄途径，她辩解到："没有文字或者铭文可以用来表达悲伤，所剩的仅仅是私人领域，个人可以在这里对此做出反应。"她继续说，"游客的自愿参加，使那些禁止缅怀的禁令被遗忘在每个人的责任感的脑后"。1

在许多纪念物中，个人标记成为最令人惊讶的现象，如姓名或者可以唤起回忆的数字。它们作为一个代表性的媒介，一种悲伤的表达方式而存在着。前面所述的特色项目展示了许多种处理这种需求的不同方法，即将个人经历带入纪念碑中，成为一个关键元素。

在纽约的911国家纪念碑中，设计师们创造了一个刻有所有受害者名字的护墙，以俯瞰在2001年臭名昭著的恐怖袭击中被损毁的两座世贸大厦的上空体量。根据该项目的创造者所述，在世人与逝人之间创造一个临界处，是一个有意识的决定。然后，这个纪念碑似乎完全承载了一个时刻，即将个人领域、名字赋予意义，以表达这一悲剧的绝对重要性。这是这一纪念碑的中心部分，甚至刻有遇难者名字的设计都服从了复杂的设计规则。纪念碑的缔造者们将其称为"有意义的邻接空间"系统，旨在使纪念碑不仅为遇难者的死亡，还为其生命所代言。

设计特点与纪念物之间的具有启发性的共鸣也是Mechelen设计的集中营纪念馆的一个重要方面，这是一座为大屠杀以及人权所建的纪念馆，博物馆和档案中心。在这个项目中，建造者利用城市原拘留所的废墟来设计一座建筑，使其能够对那些在二战期间被驱逐出城市Dossin军营，赶至德国和波兰的人们致敬，设计充分利用了这一人类悲剧的重要

Individual Experience and Collective Memory

The Holocaust is arguably the greatest tragedy that still pervades contemporaneous collective memory. Numerous memorials were dedicated to it, with some of them featured ahead in this issue. One of the most discussed was Peter Eisenman's Denkmal für die ermordeten Judsen Europas (Memorial to the Murdered Jews of Europe), inaugurated in Berlin in 2005. As Esther da Costa Meyer insightfully points out, Eisenman's design redefines the role of the memorial as "a medium capable of evoking, rather than representing, an event incommensurate with individual experience." There is no therapeutic catharsis in this memorial, she argues, *"no words or inscriptions give voice to grief, which is left to the private realm of individual response."* And she goes on contending that it is through the willed participation of the visitors that *"the injunction to remember is left to the responsibility of each individual spectator."* [1]
In many memorials, however, one of the most striking phenomena is the use of tokens of the individual, such as names or evocative numbers, as a representational medium, an expression of grief. The projects featured ahead suggest many different ways to tackle this need to bring about individual experience as a crucial component of a memorial.
In New York's National September 11 Memorial, the designers created a parapet with the names of all victims, overlooking the voids of the two WTC towers destroyed in the infamous attacks of 2001. According to the author of the project, this is a conscious decision of creating a threshold dividing the living and the dead. It seems, however, as if this utterly charged moment could only make sense bringing about the realm of the individual, its name, to cope with the overwhelming magnitude of the tragedy. This was such a central aspect of the memorial's design that even the placement of the victims' names obeyed to rather complex criteria. The authors called it a system of "meaningful adjacencies" aimed at making the memorial speak "not only of the victims' deaths but also of their lives."
The suggestive resonance between design features and the object of memorialization is also an important aspect of Mechelen's Kazerne Dossin, a memorial, museum and documentation center on Holocaust and human rights. In this project the authors used the ruins of the city's former house of detention to design a building that should pay tribute to those people that were deported

废奴纪念馆，位于卢瓦尔河前侧的滨海大道内
Memorial to the Abolition of Slavery, located in public esplanade in front of the Loire River

方面，以作为设计灵感来源。十二个铸铁支柱立在一层，作为犹太教的十二个部落的标志，而二层和三层的窗户则采用了25 267块砖进行围砌，而这个数字正是被驱逐出Dossin军营的人们的数量。

而建在波兰城市帕尔米拉的博物馆，类似于一座纪念碑，这是纪念二战期间又一惨剧的实例。两千多波兰市民被纳粹主义于帕尔米拉森林中杀害，建筑位于一个可以俯瞰公墓区的场地，而在这片公墓区内埋葬着2252位遇难者。作为这些遇难者的栖息地，这座建筑的外墙覆以生锈的钢板，钢板上的穿孔为子弹大小，穿透了建筑的表皮，提醒着人们帕尔米拉悲剧中遇难者的数目。其设计者认为，这一理念旨在创造一座可以缅怀的建筑。

在上面讨论的案例中，有一个个案较为引人注目，即每一个人都采用一个独特的标志来代表：一个名字、一块大理石板、一块砖、一个柱形物，或者是墙身的一个洞。这一举动见证了在一座纪念物中理解个人悲伤过程中产生了无法抵抗的困难，而建筑物也成为大家集体缅怀的成果。

空间隐喻和身临其境的氛围

采用模拟的策略，则是另一种应对设计纪念物的挑战。为了模仿或者模拟纪念事件，其发生的场地通常被用来激发设计师的设计手法。在废奴纪念馆（建在法国南特市）中，设计团队利用卢瓦尔河畔的地下居住空间来创造，用他们的话来说，"一个为废除奴隶制的斗争而进行的挖掘行动，带有一定的隐喻性和情感性"。

他们设计了一系列暴露在外的结构（暴露在这个城市及其开放的公共领域）以及身临其境的空间（即装有照明装饰的地下空间，用于展览无数次奴隶贸易中驶离南特市的奴隶船的甲板）。在这个项目中，场地的空间特点嵌入在设计中，以产生一个空间隐喻，而这个隐喻能够成功地创造一个缅怀和纪念奴隶废除的空间。然而，最重要的是，要使现在与过去重逢，使人们提高奴隶问题的意识。

类似的设计方法在Zanis Lipke纪念馆中也有所体现。纳粹占领拉脱维亚期间，在Zanis先生和Lipke夫人隐藏了50多名犹太人的地方，设计

from the city's Dossin Barracks to concentration camps in Germany and Poland during World War II. Important aspects of that human tragedy were used as a source of inspiration for the design of the building. Twelve cast-iron structural columns were showed on the ground floor as a token of the twelve tribes of Judaism, and the windows of the second and third floors were closed up with a total of 25,267 bricks, which was exactly the number of persons deported from the Dossin barracks.

The museum built in Poland's city of Palmiry, was created as a memorial to yet another instance of the horror created by World War II: the more than two thousand Polish civilians murdered in Palmiry woods by the Nazis. The building is located in a position where it overlooks the cemetery where each of the 2,252 memorialized victims are buried. As homage to each of these victims, the external wall of the building is cladded with rusted steel panels perforated with bullet-sized holes piercing its skin and evoking the number of victims of Palmiry's tragedy. The idea, the authors argue, "was to create an architecture of remembrance."

In the cases discussed above there is a conspicuous presence of the individual, each and every one is represented by a distinctive token: a name, a marble slab, a brick, a column, a hole in the wall. This bears witness to the overwhelming difficulty in making sense of the individual grief in a memorial that should also be a collective endeavor of remembrance.

Spatial Metaphors and Immersive Atmospheres

A strategy of simulation is another way to cope with the challenge of designing a memorial. To emulate or simulate the context in which the object of memorialization can be experienced is often used to spark the designer's approach. In the Memorial to the Abolition of Slavery, built in the French city of Nantes, the team of designers used an underground residual space in the riverfront of the Loire River to create, in their words, "a metaphorical and emotional evocation of the struggle for the abolition of slavery". They devised a sequence of exposure – to the city and its open public realm – and immersion – to an ill illuminated underground space that suggests the slaves' deck of the ships that departed from Nantes in countless slave trading expeditions. In this project, the spatial characteristics of the site were embedded in the design to produce a spatial metaphor that succeeds in creating spaces for remembrance and commemoration (the abolition of slavery), but, above all, confronts the present with the past and

1. Esther da Costa Meyer, "Speak, Memory. On Peter Eisenman's Holocaust Memorial", Artforum 44, vol.5, 2006, pp.47~48.
2. Paul Ricoeur, *Memory, History, Forgetting*, trans. Kathleen Blamey and David Pellauer (University Of Chicago Press, 2006).

集中营纪念馆利用之前的拘留所来纪念二战期间被驱逐出Dossin军营的人们
Kazerne Dossin used former house of detention to pay tribute to people deported from Dossin Barracks during World War II

师们创造了一座隐喻的建筑，这座建筑能够与当地的建筑（即典型的谷仓）以及倾倒的轮船产生共鸣。据设计师介绍，这座建筑代表了在洪水泛滥的港口中幸存的人们的生活。这个项目由一个长廊建筑所定义，而这个长廊被过度地利用，来创造一个空间序列，使来访者能够面对阴暗与明亮、水平位与垂直位、地面与天空之间的强烈对比。总的来说，使他们面对地狱与天堂画面之间的强烈对比。

在上述的大多数项目中，我们可以注意到一种建筑手法，这种手法绝对是由身临其境的氛围所定义的。在大多数案例中，它能够使来访者在这个感知空间内产生焦虑。而这个纪念物便成为有形的物体，之后，这些空间便会产生类似神圣的特点，而那些惨痛事件所遗留下来的遗迹则会有助于产生一种接近宗教性质的氛围。而在少数的几个案例中，一个纪念物则是一处由专门拨款修建的空间或开放式的立面，如柏林的Eisenman纪念碑。事实上，这种设计手法是一个矛盾体：宝贵的纪念似乎与氛围轻松的拨款或者建造相悖。虽然如此，在墨西哥暴乱遇难者纪念馆中，其缔造者则在设计一座纪念物方面取得了成功，这座建筑既创造了一个充满物质隐喻的氛围，使人们身临其境，同时其本身也是一处真正的公共空间，呈现出开放式的转变。如作者所说，"一个项目，应该面向城市以及市民捐赠的款项所开放"。自然和建筑之间的有形互动成为抵制暴力及与之相关的破坏的一剂良药。设计师设计的这个项目可以被看做是与Chapultepec公园大型绿地相对应的70面金属墙，利用不同类型的钢铁成的构件来展示其与公园游客之间可能进行的大量互动。

在纪念碑的设计中，如同前述的几个特色项目，建筑师面临着要超越学科问题的挑战。记忆的概念，如保罗·利科所建议的，要阐释什么是应该被纪念的，什么应该被忘记，以及什么应该被原谅。[2]在这一复杂的探讨中，纪念物的设计师们必须仔细思考，将所缺失的东西呈现在大家眼前。

raises consciousness on the issue of slavery.
A similar approach can be seen in the Zanis Lipke Memorial. In the place where Mr. Zanis and Mrs. Lipke hid more than fifty Jews during the Nazi occupation of Latvia, the designers created a metaphorical form that resonates both with vernacular constructions (the typical barns of the area) and an overturned ship, representing, according to the authors, "the post-flood arch harboring's surviving life". The project is defined by a promenade architecturale that is overtly designed to create a spatial sequence that confronts the visitor with harsh transitions between obscurity and light, horizontal and vertical, ground and sky; in short, between images of hell and heaven.
In most of the projects discussed above, one can observe an architectural approach that is strongly determined by the creation of immersive atmospheres. In most cases, that is deliberately meant to create some un-easiness for the visitor in the perception of space. The object of memorialization becomes almost tangible; these spaces then acquire a quasi-sacred character, where relics of the traumatic event contribute to creating an almost religious experience. In few cases, however, a memorial is a space or a surface open for appropriation as suggested in Eisenman's memorial for Berlin. In fact, this almost resonates with a paradox: valuing memories is seemingly at odds with light-hearted appropriations and manipulations. Nevertheless, in the case of the Memorial to Victims of Violence in Mexico, the authors succeeded in designing a memorial that creates both an immersive atmosphere full of material metaphors and a true public space open for transformation. A project, as the authors claim, "opens to the city and opens to the appropriation by the citizens". The tangible interaction between nature and construction becomes an antidote against violence, and the destruction associated with it. The authors designed the memorial as seventy metallic walls set against the lavish greenery of Chapultepec Park, using elements made of different types of steel to suggest a multitude of possible interactions with the users of the park.
In the design of memorials, as can be seen in the projects featured ahead, the architects face challenges that go beyond mere disciplinary issues. The notion of memory, as Paul Ricoeur suggested, deals with what ought to be remembered, what can be forgotten, and what might be forgiven.[2] It is in this complex negotiation that the designer of a memorial has to dwell, bringing about the presence of something absent. Nelson Mota

该工程建在南特市,作为曾经的奴隶贩卖中心,这个地点本身就是"见证者"。其设计体现出两个基本的姿态:揭露和沉浸。二者共同创造出富有层次的深度体验,游客可以发现并理解历史和地址所唤醒的维度空间。这座建筑没有设置直接的标志,而是试图让游客寻找记忆,使打开空间就可以发现内在的记忆的这一理念更加清晰。

这一项目把原来用于泊车的旧码头地区改建为意义重大的公共空间及值得纪念的场地。项目将重点放在把原来的地下空间(18、19、20世纪卢瓦尔河的港口及河堤墙遗址)改建为"通道",即一个地下纪念空间。而这处被"发现"于档案资料中的地下空间,沿着河流甚至低于水位建造,是纪念馆的中心。由于卢瓦尔河的潮水每天都高于4m,将"发现的空间"改造为公共空间需要非常复杂的工程,以便建造具有保护性的基础墙面防水层(原有结构下的堤坝)。

废奴纪念馆

Wodiczko + Bonder

Pont Anne de Bretagne和Passarelle Victor Schoelcher之间的纪念场地处嵌入了2000块玻璃,唤起人们对庞大的奴隶交易数目以及运奴船的记忆。

游客可以在平坦的公共空地处进入"通道"(两边均可进入),然后发现自己身处在19世纪就存在的河堤墙及20世纪的河堤墙混凝土结构围合起来的狭长空间内。纪念碑通过硕大的倾斜玻璃嵌板对废奴制进行庆祝,在南特市犹如同一只铁铲将土壤切割开来。这面玻璃墙上精选了各大洲受奴隶制度的触动所发布的文本以及贩卖人口的形式,时间横跨五个世纪(从17世纪到21世纪)的斗争历程。

由于通道低于城市地平线约5m的高度,因此建筑师还设计了电梯、专用照明及其他便于进入的元素和通用的设备。

纪念馆设置了信息区,让游客了解关于奴隶制及横跨大西洋的奴隶交易等历史事实。针对各种形式的奴隶制及奴隶贩卖的当代斗争情况也通过一个类似于"情景室"的空间揭露出来,这个房间面对立法大楼(由让·努维尔设计)。目前人们设想建造一条1.5公里长的全新城市场地来作为一种多层次的城市设计方式,以此来建造记忆和城市,包括Chateau des Ducs de Bretagne(在这里大量展览室用来呈现南特作为奴隶贩卖港口的历史)、公共通道(展现南特及奴隶制度,包括参观奴隶房屋、奴隶商船旅馆)和卢瓦尔河纪念馆。

除了象征性的意义,纪念馆还用于宣誓场地及南特市一年两次的人权论坛的特别会议地址,以此来加强这一地址用于纪念和展现废奴制度的特殊性。这表明当代的解放运动依然英勇无畏地持续进行,提醒人们于1848年达到高潮的废奴运动绝非徒劳,希望有朝一日废奴工作将全无必要。这处新的公共空间以及城市景观作为一个涉及政治、城市、艺术景观以及建筑的项目,解释了南特及世界各地废奴运动的艰难过去和当今情况,并且希望成为转变行为、人权行为主义以及公民行为的代理和催化剂。

Memorial to the Abolition of Slavery

The project is based on the site and situation of Nantes, one that already speaks of and for itself as a "witness". Its design proceeds through two fundamental gestures, exposure and immersion, which together create a layered, in-depth experience through which visitors may discover and interpret the evoked dimensions of history and a place. The architect attempted to engage the visitor in the search for memory, through absences of direct signs, making the idea of opening up spaces for memory to be found within palpable.

照片提供; courtesy of the architect

This project transforms the site, the old port areas, from current uses such as parking, into a significant public space and commemorative ground. The project is centered on adapting a sub-surface preexisting space (residue of construction of the ports and embankments walls of the Loire in 18th, 19th, 20th Century) into a "passage", an underground memorial space. The sub-surface space, "found" in archival documentation, along and often below the level of the water, is the heart of the memorial. The transformation of this "found space" into public space required very complex engineering in order to construct a protective "cuvelage" (a dam under the old structures) due to the tides of Loire River (more than 4m daily).

Between Pont Anne de Bretagne and Passarelle Victor Schoelcher, 2,000 glass inserts, evoking the magnitude of the slave trade and the memory of slave-vessels, were set into the commemorative ground.

From the public esplanade, visitors can access the "passage", on either end, to find themselves in a long space enclosed by the pre-existing 19th century embankment wall and the 20th century embankment concrete structures. The monument celebrates the great rupture of abolition with the thrust of a great slanted glass plate, which emerges to the city as a spade that cuts the soil. On this glass wall selected texts from different continents touched by slavery and forms of human trafficking and from periods spanning 5 centuries of struggle (XVII Century to XXI Century) can be found. An elevator and special lighting and other elements for accessible

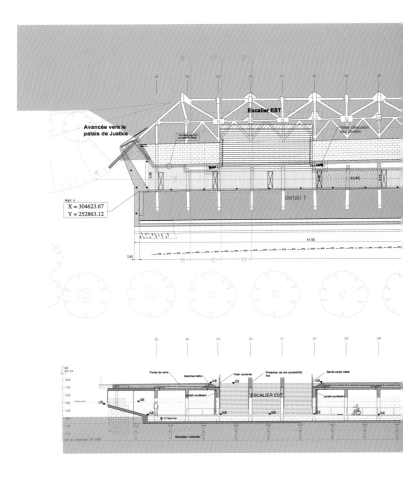

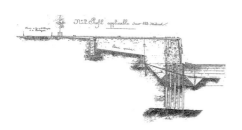

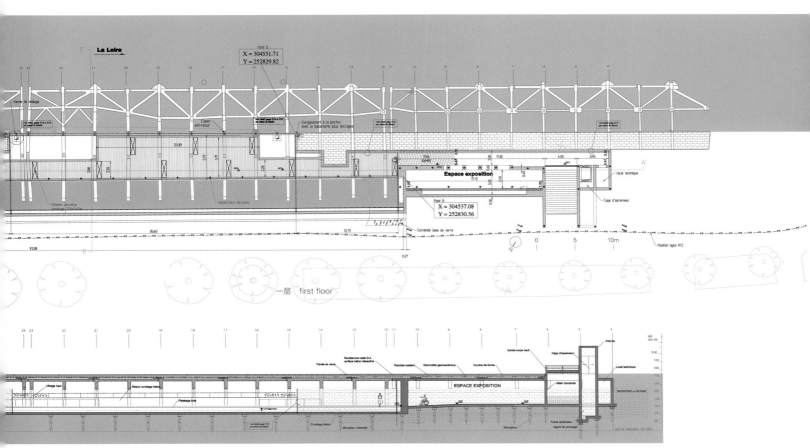

一层 first floor

A-A' 剖面图　section A-A'

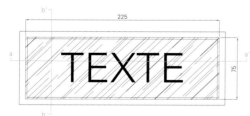

1. produit verrier
2. protection durant désactivation béton
3. béton désactivé
4. béton
5. mastic structurel
6. mastic elastomére 1er cat.
7. acier inox 304L
8. callage
9. ciment de finition (lot beton)

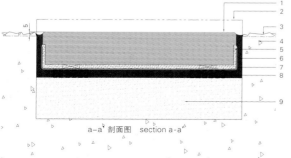

a-a' 剖面图   section a-a'

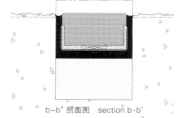

b-b' 剖面图   section b-b'

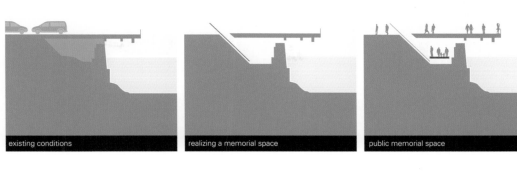

existing conditions | realizing a memorial space | public memorial space

flow   ebb

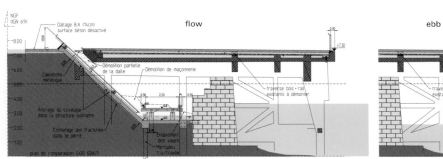

B-B' 剖面图　section B-B'

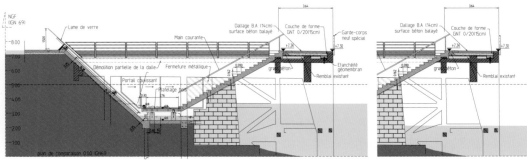

C-C' 剖面图　section C-C'

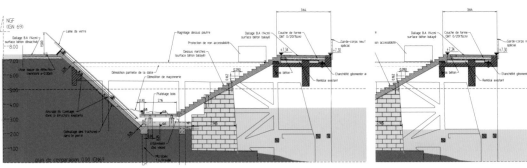

D-D' 剖面图　section D-D'

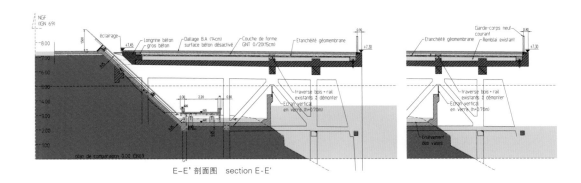

E-E' 剖面图　section E-E'

and universal design are included, as the passage is about 5m below the level of the city.

The memorial includes an information area where visitors orient themselves about historical facts about slavery and the transatlantic trade. The dimensions of present-day struggles against all forms of slavery or the slave trade will be exposed, within a space acting like a "situation room", facing the Palais de Justice (Jean Nouvel). As part of a multilayered urban approach to memory and the city, a new 1.5km-long urban parcours has been conceived. It includes the Chateau des Ducs de Bretagne (where a number of exhibition rooms are dedicated to present the history of Nantes as a Slave Trading Port), a public promenade (Nantes and Slavery, including houses to visit and hotels of Ship/Slave Merchants), and the Memorial on the Loire.

Beyond its symbolic purpose, the memorial will be used as a space for testimonies and as a special meeting site during the bi-annual human-rights forum to be held in Nantes. This would thus confirm the specificity of this site as a responsive site of memory and of struggle. Signaling the contemporary work of liberation that courageously goes on will serve to remind us that the work of abolition that culminated in 1848 was not in vain, and that perhaps its work will one day hopefully be no longer necessary. By shedding light over difficult pasts and presents, both in Nantes and the world, and as a political, urban, artistic, landscape and architectural project, this new public space and urban landscape, hopes to become an agent and catalyst for transformative action, human rights activism, and civic engagement.

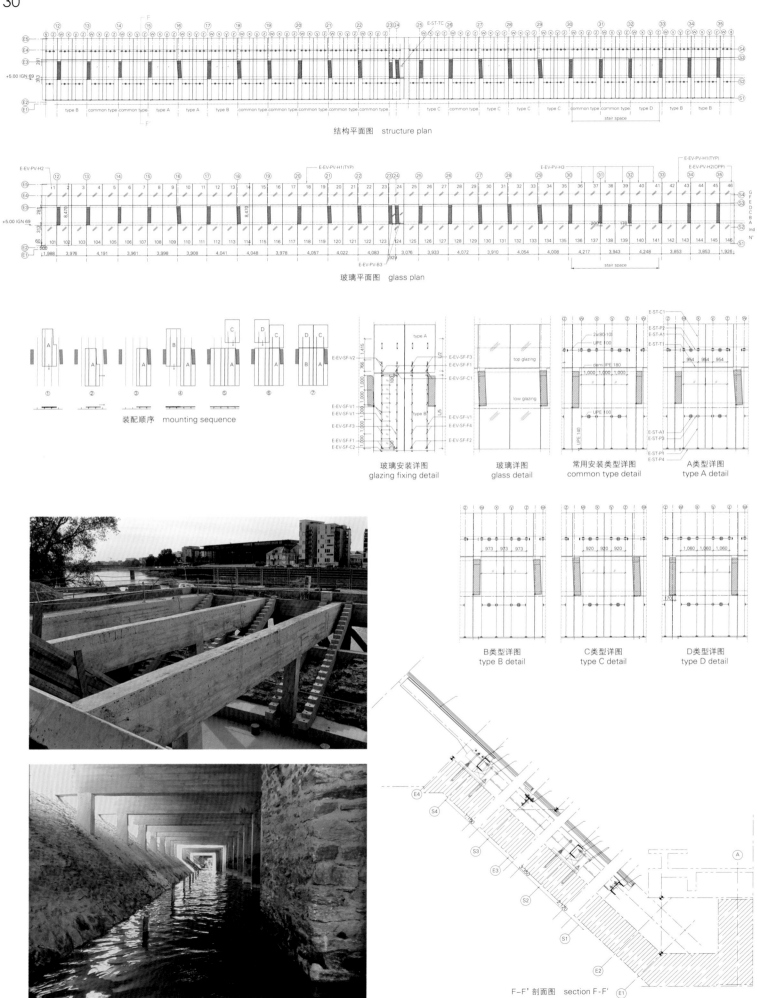

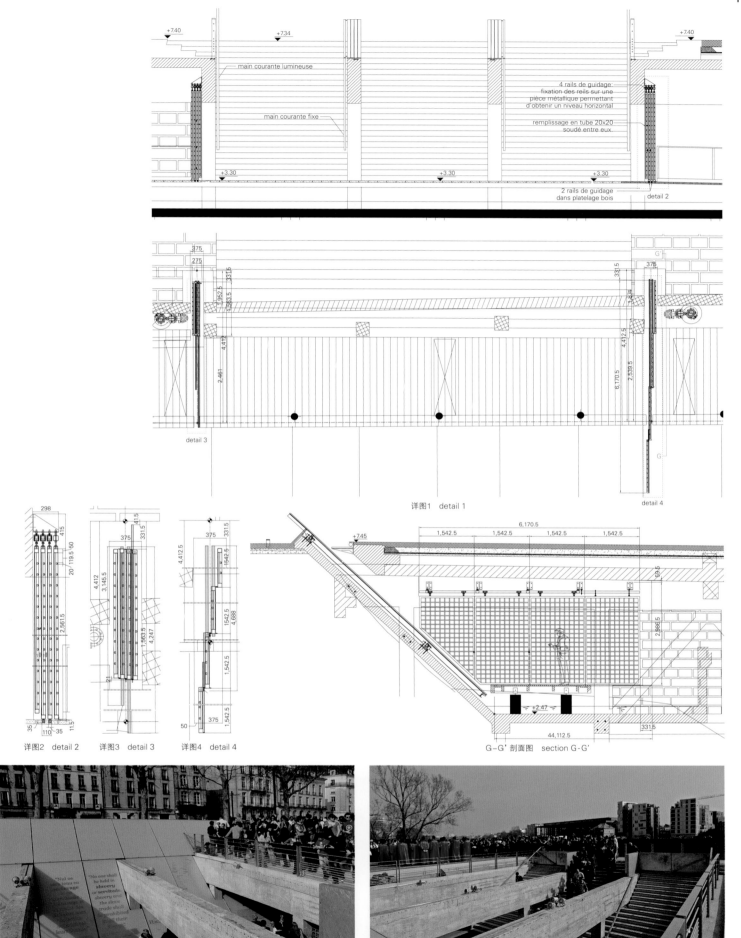

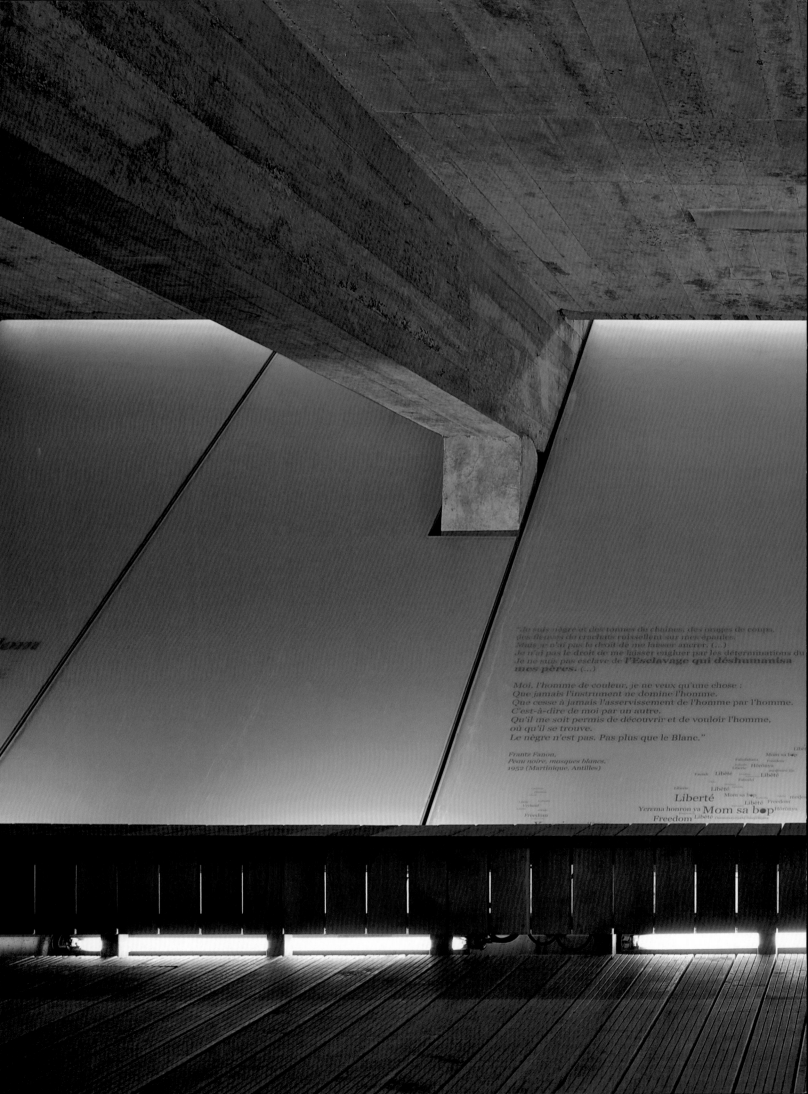

项目名称：Memorial De L'abolition De L'esclavage
地点：Nantes, France
建筑师：Krzysztof Wodiczko, Julian Bonder
合作商：Thomas Long, Snehal Intwala, Nicholas Capone, James Shen, Ryan McClain, Bill Panasuik, Patrick Charles, Maximo Rohm, Ron Henderson, Michael Blier, Martial Huguin, Emmanuelle Chérel
玻璃顾问：James Carpenter Design Associates, RFR Engineers, Philippe Bompas & Niccolo Baldassini
总承建人：DLE Ouest, Nicholas Boterf
玻璃墙体设计：Polar, Torino, Paolo Cherasco
照明设计：Citelum
景观建设：ISIS Espaces Vertes
公共空间面积：8,500m² 总建筑面积：1,500m² 竣工时间：2011
摄影师：©Philippe Ruault (courtesy of the architect) (except as noted)

911国家纪念馆

Handel Architects

911国家纪念馆占地32 375m², 位于人口稠密的下曼哈顿城区, 该广场原址为世贸双塔。纪念馆广场成为重建的世贸中心综合体（面积为64 750m²）的一部分, 将场地和城市较远处的地方连接在一起。开放友好的设计培养公众集会的民主价值, 这在2001年9月11日恐怖袭击后纽约城的集体反应中起到了关键作用。

纪念馆广场在市中心形成了32 375m²的空地, 近400株沼泽白橡树形成了渗透光线的棚顶, 游客来到纪念馆会自觉地向中心行走, 然后看到两个反射水池, 深深地"刺破"了广场的广阔空间, 形成了空空的容器, 两处池塘都凹入地面9m深, 且四周围以瀑布, 以勾画出双塔原址的轮廓, 上空体量尽管无形却如同存在一般, 人们可以想像得到原址的模样。

每块上空体量的周围是2.4m宽, 0.6m高的泻水台, 这里不仅是水源还是人工瀑布的喷水孔, 泻水台及上空体量都由弗吉尼亚州的深灰色花岗石所覆盖。水渠边沿为金属锯齿边缘, 引导水流到不同的通道, 成千股水流形成翻腾的水幕, 使数千个生命的逝去及集体丧生之痛萦绕心头。

散发着幽暗光泽的黄铜板上篆刻着遇难者的姓名, 空缺的材料在白天使这些名字形成阴影。晚上铜板盘旋如翼的侧面从内部照亮, 每个名字以柔和的光芒相互辉映。铜板外观的设计简单, 但是在其他方面极为复杂, 经过精心设计, 加热、冷却、灯光及热膨胀机理完全嵌入铜板内部, 遇难者名单看似无序, 实则精心安排, 每个遇难者的家庭都受邀参加设计程序, 建议选择和哪个受害者的名字相毗邻。设计者询问了1200多名家属, 用接近一年的时间以高强度的设计完成了看似不可能的任务。最后设计满足了每个家属的要求, 纪念碑上的每个名字的位置既独特又具个人色彩。

该设计基础的核心指导原则之一是相信公共空间的恢复能力, 以及民主社会中作为公众集会容器的重要的公民功能。设计源自城市个体目击者的第一手灼痛的经验, 他们作为一个群体到此集会, 彼此相互支持, 以勇气、同情和坚韧来面对野蛮行径。这一经验引导建筑师坚定地相信这一场址应当再次和城市生活相联系, 成为日常存在的一部分。在这个场地, 庄严的凝视可以与工作及娱乐的上班族和居民紧密联系起来。这也是一处过去和现在交织的地方。

National September 11 Memorial

The National September 11 Memorial is an eight acre plaza set within the dense urban fabric of Lower Manhattan, where the former World Trade Center Twin Towers once stood. The Memorial Plaza is an integral part of the sixteen-acre redeveloped World Trade Center Complex, and it reaches and connects the site to the city beyond. It is an open and welcoming design that is meant to foster the democratic values of public assembly that played such a pivotal role in New York City's collective response to the attacks of September 11, 2001.

The Memorial Plaza forms an eight acre clearing in the middle of the city and is vaulted by a permeable canopy of close to four hundred swamp white oak trees. As visitors to the memorial make their way towards the center of this space, they encounter two reflecting pools that deeply puncture the vast flat expanse of the plaza, and form empty vessels. They are recessed thirty feet into the ground and are lined by waterfalls, delineating the location of the former towers. The voids are absence but made present and visible.

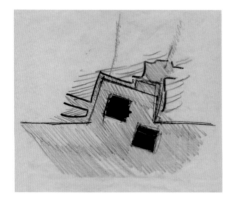

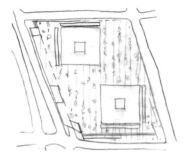

ABACUS-LIKE BANDS

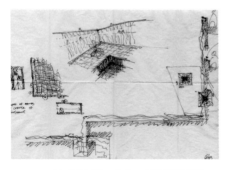

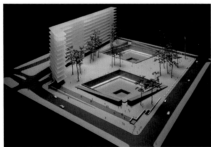

Surrounding each acre-sized void is an eight-foot wide and two-foot high water table that serves as the source and springing point of these waterfalls. The water table and the voids are clad in a dark gray granite that was quarried in Virginia. The serrated metal edge of the weir channels the water into separate streams and evokes the haunting loss of thousands of individual lives and the collective loss suffered by all with its billowing curtain of water composed of thousands of individual strands of water.

The names of the victims are incised into darkly patinated bronze panels and appear as shadows during the day, marked by the absence of material. At night, the hovering wing-like profile of the panels is illuminated from within, lighting each name with a soft glow. The panels are simple in appearance but are complex in every other aspect, with precisely engineered and completely concealed heating, cooling, lighting and thermal expansion mechanisms. The names that appear to be in no discernible order are in fact carefully composed in an arrangement that emerged when every family of a victim was asked to participate in the design process by suggesting what other victims' names should be placed adjacent to the name of the person that they lost. Over twelve hundred individual requests were made and close to one year of intense design work was required to resolve what seemed like an impossible task. The final arrangement that emerged met each and every request and placed each name in a physical location in the memorial that is unique and personal.

One of the core guiding principles that served as a foundation to this design was the belief in the resiliency of public space and its important civic function as a vessel for assembly by people in a democracy. This design sprang from the first hand and searing experience of witnessing individuals in a city coming together as a community in its public spaces to support one another and look savagery in the eye with courage, compassion and stoicism. That experience led the architects to firmly believe that this site should be reconnected to the life of the city and made part of its everyday existence. It is a place where solemn contemplation can and should rub shoulders with office workers and residents who are at work and play. It is a place where the past and the present come together.

项目名称：National September 11 Memorial
地点：New York City, NY, USA
建筑师：Handel Architects LLP
设计师：Michael Arad
建筑设计师：Michael Arad, Gary Handel, Amanda Sachs, David Margolis, Robert Jamieson, Cristobal Canas, Garrett Brignoli
合作建筑师：Davis Brody Bond Aedas
景观建筑师：Peter Walker and Partners
喷泉设计师：DEW Inc.
施工经理：Lend Lease Inc.
照明设计师：Fisher Marantz Stone
结构工程师：WSP Cantor Seinuk
MEP工程师：Jaros Baum & Bolles Consulting Engineers
地质技术工程师：Mueser Rutledge Consulting Engineers, Langan Engineering & Environmental Services
安全工程师：ARUP
防水：WJE Engineers & Architects P.C.
工程师：Simpson Gumpertz & Heger Inc.
可持续性顾问：Viridian Energy & Environmental LLC
排气&硬化顾问：Weidlinger Consulting Engineers
委托人：Horizon Engineers
编码顾问：Code Consultants, Inc.
甲方：The Port Authority of New York and New Jersey
用地面积：32,374.85m²
竣工时间：2012.9
摄影师：courtesy of the architect-p.137, p.138~139, p.169 bottom left, bottom right, p.143 bottom left, right, ©Joe Woolhead (courtesy of the architect)-p.141, p.142 top, p.143 top

墨西哥暴乱遇难者纪念馆

Gaeta-Springall Arquitectos

公园位于墨西哥城最重要的场地Chapultepec，这部分森林属于联邦政府，多年来由墨西哥国防部监管，这个纪念馆项目旨在将公共空间从一片废墟中复原。

人们在纪念馆建筑中能够发现文化和历史的纪念和记忆，尤其是在墨西哥暴乱遇难者纪念馆中，建筑师可以从建筑角度将墨西哥社会最重要和最当前的事件之———暴乱具体化。这是个巨大的裸露伤口，针对伤口，建筑师提出了在场地上建立开放的项目，对城市开放，对市民捐赠的经费开放；这个项目是城市及其创造者的强壮纽带，其核心是公众空间的复原能力和对暴乱受害者的缅怀。

这个项目承载着双重环境的功能，它既是公共空间，又是纪念馆。第一个前提是承认场地具有森林的功能，要非常强烈地体现自然，场地的树木就体现了这一特点。

暴乱通过两个维度来体现：空间及建筑。项目所规划的上空空间是钢质墙壁及树木所形成的空间，这部分空间或是空出的地方可以提醒人们纪念不存在的或毫无人迹的理念，钢质墙壁的表面，无论是锈迹斑斑还是明镜可鉴，都展示了人类迷失自我，增值自我或是使自我朝多元化发展的特点。

大型反射由建造的70面考顿钢制成的墙体形成，它们在树林间拔地而起，在自然和建筑间扮演双重角色：树木之林及铁墙之林。树木群落及游客扮演生物，墙体群落则发挥着具有非物质性的遇难者记忆的角色。

材质的清单有所缩减：钢铁、混凝土以及森林的自然元素。建筑师使用三种考顿钢：自然型、生锈型以及不锈钢镜面型，每种钢代表不同的涵义。生锈型钢表明人们一生中的标记和伤痕。不锈钢镜面型则用来反映生命，并使生命多元化：人类、树木以及中央空间的水，自然型钢用来表示不受扰乱的元素，提醒人们社会必须保持稳定、主要的及至关重要的价值。混凝土建造的小路和长凳，可以用来行走和反思。

在中央空间，也就是纪念馆的主要空间，有一个1200m²的喷泉，形状不确定，外形较为开放，提醒人们暴乱事件依然没有停止。喷泉用隔栅覆盖，游客可以行走在水面上。水意味着生命，水清洗污垢，水愈合伤痛。

在这片区域，钢质墙壁更加坚固、也更加高大，创造出整个场地的最突出的部分，水面上树木和墙体的倒影吸引人们的注意，使人们上下观望。当他们向上望去时，可以看见蓝天、光线、太阳以及希望。

最后，项目最重要的一部分是钢质墙壁的人性化特点和比例。社会负责建造纪念馆，70面金属墙是人们书写遇难者名字的空间，表达出人们的痛苦、愤怒和渴望。这些钢质墙壁的功能犹如镜子和黑板，以文字的形式将有组织的暴力犯罪转换为疼痛和毁灭的目击者。

Memorial to Victims of Violence in Mexico

The park is in Chapultepec, the most important site of Mexico City. This part of the forest belongs to the Federal Government and was under the custody of the Ministry of Defense of Mexico for many decades. The purpose of the memorial project was to recuperate from ruined site in terms of public space.

A memorial is the architectural piece in which the architects can find the remembrance and the memory of the culture and history; in the particular case of the Memorial to Victims of the Violence in Mexico, the architects materialize, in terms of architecture, one of the most important and current issues of Mexican society: violence. This is the big and open wound; in response to this, they propose an open project in the site, open to the city and open to the appropriation by the citizens, a project with a strong relationship with the city and her actors. The recuperation of the public space and the remembrance of the victims of violence are the essence of the project.

The project plays the double condition of public space and memorial. The first premise was to recognize the vocation of the site

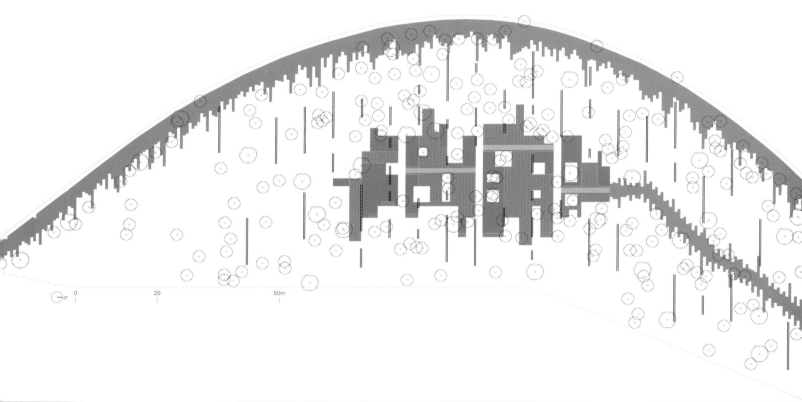

项目名称：Memorial to Victims of Violence in Mexico
地点：Reforma Avenue, Chapultepec, Mexico City, Mexico
建筑师：Julio Gaeta, Luby Springall, Ricardo López
用地面积：15,000 m²
造价：USD 2,550,000
竣工时间：2013
摄影师：courtesy of the architect - p.147 bottom, p.148~149, p.151, p.153 top,
©Sandra Pereznieto (courtesy of the architect) - p.150, p.152,
©Christian Ortega (courtesy of the architect) - p.145,
©Daniel James (courtesy of the architect) - p.153 bottom

as a forest; with a very strong presence of nature; the trees are there and they characterize the site.

The violence is suggested in two dimensions: the void and the built. The void proposed in the project is the space created between the steel walls and the trees. This void or empty space could remind us the concept of the no-presence and absence of the people to remember, and the surfaces of the steel walls, rusty or mirroring, show that we can lose ourselves, add ourselves, or multiply ourselves.

The big projectual action consists on building seventy metallic walls in corten steel rising between the trees; it is a dual play between nature and architecture: the forest of trees and the forest of walls. The society of the trees and the visitors play the living beings; the society of the walls play the unmaterial of the memories of the victims.

The list of materials is reduced: steel and concrete, added to the natural elements of the forest. The architects are using the corten steel in three ways: natural, rusty or stainless mirroring, each of them with different meanings. The rusty steel means the marks and scars that time makes in our lifetime. The stainless mirroring steel is used to reflect and multiply the living: persons, trees, and the water of the central space; and the natural steel is used as an unperturbed element that reminds people the main and essential values that societies must keep to live in peace. Concrete is used for the lanes and the benches; for walking and reflection.

In the central space, which is the main space of the memorial, there is a 1,200m² fountain with an undetermined form and open geometry, to remind people that the violence issue is still opened. The fountain is covered with a grid so that the visitor can walk over the water. Water means life; water cleans, and water heals.

In this area the steel walls rise stronger and taller, creating the strongest drama in the whole place. The reflection in the water of trees and walls make people's eyes go up and down. When they go up, they see the sky, the light, the sun, the hope.

Finally, one of the most important parts of the project is the humanization and appropriation of the steel walls. Society is responsible for making the Memorial. The seventy metallic walls are spaces for people to write the name of their victims, and express their pain, anger, and longings. These steel walls play as mirrors and blackboards, and by the writings, are transformed into witnesses of the pain and destruction provoked by the violence of the organized crime. Gaeta-Springall Arquitecto

集中营纪念馆

大屠杀及人权纪念碑、博物馆及文献中心
awg Architecten

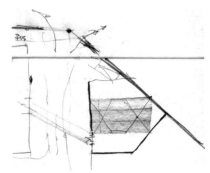

二战期间，25 267人从Dossin军营被驱逐出境，而到德国和波兰的集中营。在20世纪70年代，该建筑被改造为公寓。

和人们所想象的相反，这座新博物馆并不在军营所在处，即发生驱逐事件的所在地，而是在军营的对面——原拘留所的位置。位于这两座建筑之间的开放广场是一个交通枢纽，连接了两座建筑。

军营和博物馆前院墙之间的开放广场已经完全铺设，因此，它成为城市里可辨认的公共场所。人们在这里首先看到的是广场，以此来感受这座城市自然而真实的地方。军营及其内部庭院恢复了它们在大街上的位置。

建筑的一层并不与城市发生互动。人们在广场的中央不容易发现不太引人瞩目的入口大厅，因为它是设在建筑的另一边的。Donald Judd的理念是正确的：对称是规则，而不对称则是例外。该项目就是一个例外，它没有试图根据"规则"来解决任何问题，入口是不对称的，更确切地说，是被"搁置"在一边的。入口没有位于军营大门的对面，也没有雨篷或遮阳篷，它几乎就是一个次入口，一个滑动门；欢迎来往的轨道车。网格，即几何结构很不简单。立面的底层网格不同于内部结构布局的网格。

建筑分为两个层次：一层和顶层，人们在此可以享受自然光；而中间三层则是无窗的。一层设有12个铸铁支柱，支撑着建筑，并且代表犹太教的12个部落（是以撒的12个儿子的后裔，以撒是先祖亚伯拉罕的儿子）。

二层和三层每个窗户的尺寸都有所不同，共用了25 267块砖来进行围砌。从游客踏进一层宽敞明亮的房间起直至到达顶层，他们都在观看展品。这是与历史的一次相遇，人们重新审视自我。这里没有位于中央的中庭，没有内置的统一体，也没有或者几乎没有自然光。博物馆的房间需要控制光线，而这些光线完全来自于透视场景。大多数犹太教堂都有窗户，通风又明亮。这座博物馆有砖砌围城的窗户，用25 267块砖阻挡了日光，见证了缺失和孤独。这样一来，上面的楼层也不与城市发生互动。只有在顶层，围绕着建筑3/4的地方是可能与外面进行交流的。在这里人们可以看见这座城市，可以看到军营、军营前面的广场，更重要的是，看到军营的内庭院。

最重要的是，新大楼渴望成为一个裸露的框架，自由地被透视场景所填充。建筑的顶端和主体之间设有楼梯和电梯。主体作为展览的自由空间，而顶层则作为其他功能组成部分。

可持续发展是该建筑作品的一个主题。其重点是长寿，因此主要强调随着时间的推移，建筑的变换和其可变性。这座建筑，这座博物馆，这个地方，必须在很长一段时间内都会深深地刻在人们记忆当中。这座建筑，这个场地，渴望成为建筑文化意义的诠释，发挥博物馆作为机构的作用。

Kazerne Dossin
Memorial, Museum and Documentation Center on Holocaust and Human Rights

From the Dossin Barracks, 25,267 people were deported to concentration camps in Germany and Poland during WWII. In the 1970's, the building was subdivided into apartments.

Contrary to what one might think, the new museum is not housed in the barracks itself, the site where the deportation occurred. Rather, the space was sought across from the barracks, on the site of the former house of detention. The open square between the two buildings, as a traffic junction, links the two buildings.

The open square between the barracks and the museum's forecourt wall has been paved over. Thus it becomes a recognizable, communal location in the city. The first thing one sees is the square. In this way, people can enjoy a normal and authentic place in the city. The barracks and their inner courtyard regain their location in the street.

The ground floor does not interact with the city. At the center of the square one finds no easily conspicuous entrance hall; this is a set off to one side. Donald Judd was right: symmetry is the rule and asymmetry the exception. This project is the exception, it doesn't try to resolve anything according to "the rules"; the entrance is not symmetrical, it is, rather, "set aside". It is not situated across from the barracks gate, it has no canopy or awning, it is almost a secondary entrance, a sliding gate: a railway car's welcome. The grid, the geometry of the structure is not simple. The underlying grid for the facade differs from the grid for the layout of the inner structure.

Two levels: the ground floor and the top floor, both enjoy natural light; the three intermediate levels are windowless. On the ground floor, 12 cast-iron columns representing the 12 tribes of Judaism (descended from the twelve sons of Isaac, the son of the patriarch Abraham) support the building.

The windows on the 2nd and 3rd floors – the dimensions of which vary per window – are bricked up with a total of 25,267 bricks. From the moment one exits the large, bright common room on the ground floor up to the point that one enters the top floor, the visitor is given over to the exhibit. The encounter is a meeting with history. One rediscovers oneself. There is no central atrium, no built continuum, no or practically no, natural light. The museum rooms necessitate controlled light, light directed entirely by the scenographer. Most synagogues have windows and are airy and filled with light. This museum has bricked-up windows and 25,267 bricks keeping daylight out, bearing witness to absence and loneliness. In this way the upper floors do not interact with the city, either. Only on the top floor is it possible to circulate outside, 3/4 of the way around the building. Here one may behold the city and more importantly the barracks, the square in front of the barracks and above all, the barracks' inner courtyard.

First and foremost, the new building desires to be a bare frame to be filled in freely by the scenography. Between the building's apex and its main body are both a staircase and an elevator. While the main body remains free for exhibition space, the top floor is devoted to other components of the program.

Sustainability is a main theme in the architectural work. The focus is on longevity and mostly this therefore means an emphasis on change and changeability over time. This building, this museum, this place, must imprint upon the memory for a very long time. The building, the site, desires to be the architectural interpretation of the cultural significance and the role of the museum as institution.

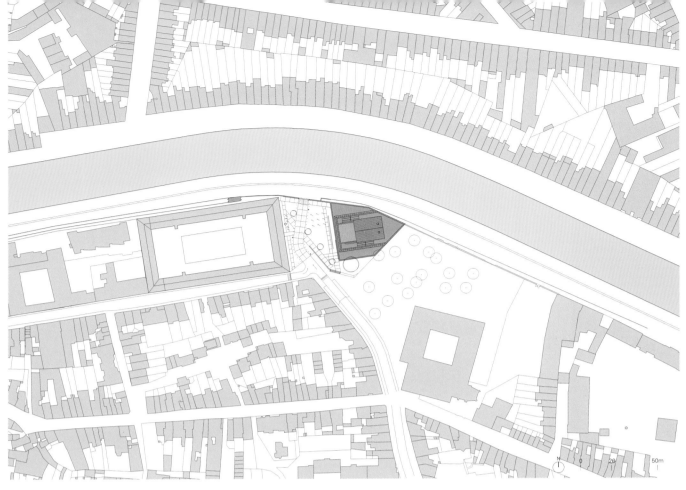

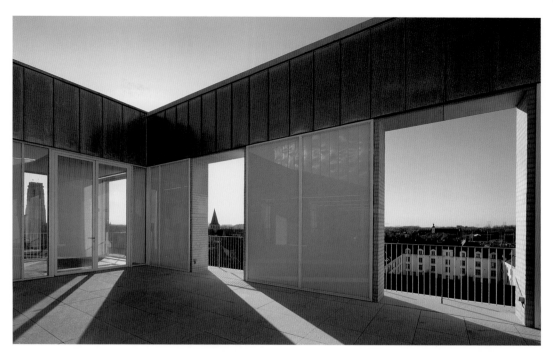

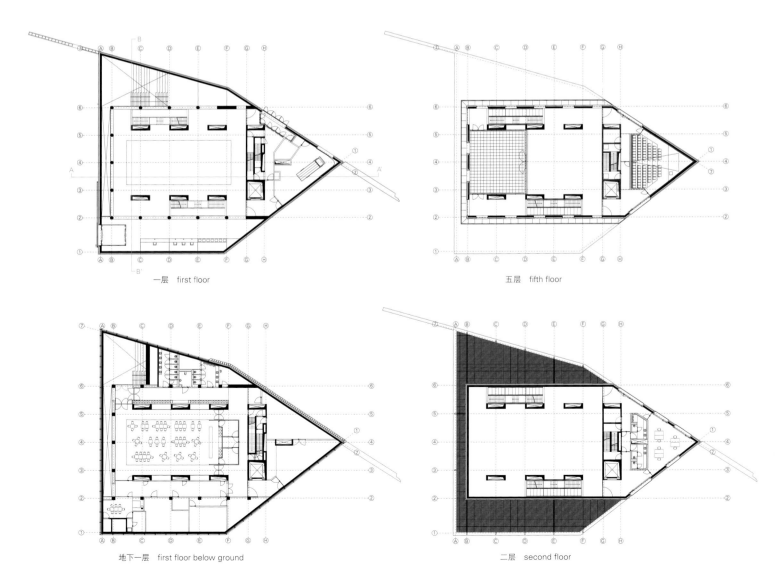

一层 first floor

五层 fifth floor

地下一层 first floor below ground

二层 second floor

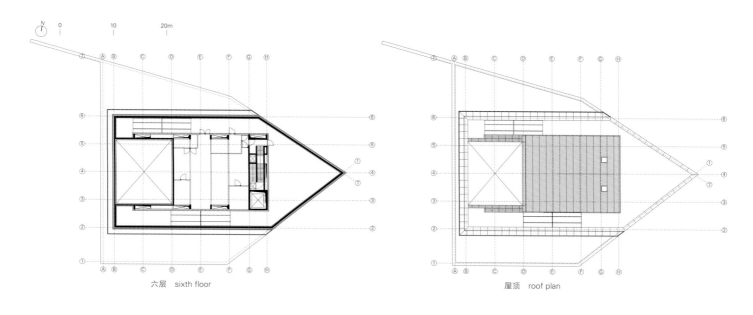

六层　sixth floor

屋顶　roof plan

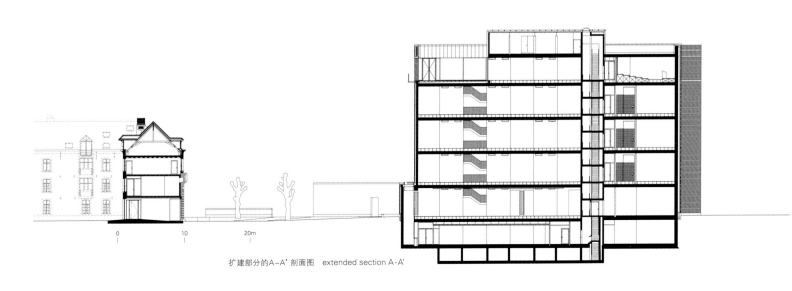

扩建部分的A-A'剖面图　extended section A-A'

项目名称：Kazerne Dossin
地点：Mechelen, Belgium
建筑师：awg Architecten
甲方：Flemish Government
用地面积：4,970m² 总建筑面积：1,235m²
有效楼层面积：6,093m²
竣工时间：2012.11
摄影师：©Stijn Bollaert - p.154~155, p.157 bottom, p.158~159, p.160, p.161 top, middle, p.162, p.163
©Christophe Ketels (courtesy of the architect) - p.156 except as noted, p.157 top, p.161 bottom, p.164, p.165

B-B' 剖面图   section B-B'

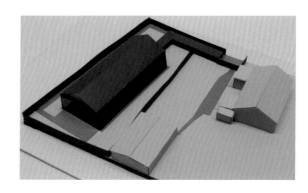

Zanis Lipke纪念馆

Zaigas Gailes Birojs

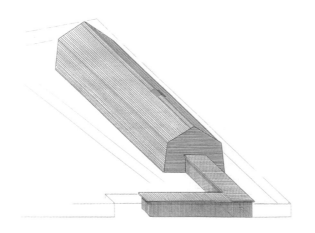

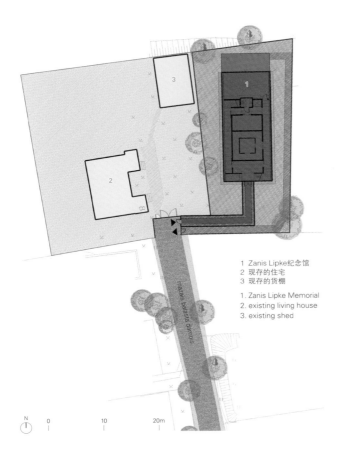

1 Zanis Lipke纪念馆
2 现存的住宅
3 现存的货棚

1. Zanis Lipke Memorial
2. existing living house
3. existing shed

拉脱维亚首都里加市中心的基普萨拉岛Maza Balasta街8号是一对拉脱维亚夫妇Zanis和Johanna Lipke拯救犹太人的地方，二战期间，他们将要被处决的犹太人藏匿于货棚下的地堡。纪念馆是一个黑色的木棚，即人们躲藏及获救的象征性避难所。建筑的形象从基普萨拉岛具有代表性的漆黑老渔民的木棚中获取灵感，博物馆从思想体系上和视觉上都和诺亚方舟以及一艘倒置的船只相像，二者都是避难的场所。

牢固的黑色木质栅栏将博物馆周围围住，封闭隧道（从巨大的入口大门开始）的通道没有启发游客意识到这座建筑真正的规模和结构，游客需要一段时间才能到达带有地下隐秘地堡的中心位置。

游客在木棚中沿着通道行走，心潮澎湃。登上阁楼，来到井边，人们通过这口井看到地下室中的地堡，这个地堡长宽高各为3m，带有木板床，还原历史上地堡的房间。阁楼是博物馆的主要展厅，稀疏的阳光从房顶射入灰尘覆盖的房间。一层还有一条迂回的路围绕着地堡。地堡上面建造了一个苏克棚，用彩绘的透明纸墙和小窗修饰成理想的家园，这是象征性的犹太教神圣的临时避难所，在残酷的世界中庇护着人类。

建筑材料：木质结构建筑，交叠的木板屋顶和建筑立面，木质彩绘的混凝土墙体，木板和潮湿抛光的混凝土地面，木板天花板。由于热保温系统长期提供低温，因此建筑为被动房。

建筑由Zanis Lipke纪念馆筹集资金协会捐赠投资。

2012年10月3日到5日在新加坡举行的世界建筑节文化类别中，Zanis Lipke纪念馆入围最后的候选文化建筑。

Zanis Lipke Memorial

Maza Balasta Street 8 in Kipsala, Riga is a place where a Latvian working class couple Zanis and Johanna Lipke saved Jews sentenced to death during the World War II by hiding them in a pit under a shed. The memorial building is the Black Shed – a symbolical shelter under which people were hidden and saved. The image of the building is inspired by old fishermen's pitch black sheds characteristic to Kipsala island. Image of the museum ideologically and visually resembles Noah's Ark and an inverted boat – both being places for shelter.

项目名称：Zanis Lipke Memorial
地点：Mazais Balasta dambis 8, Riga, Latvia
建筑师：Zaiga Gailes, Ingmars Atavs, Agnese Sirma, Ineta Solzemniece, Zane Dzintara
理念设计师：Maris Gailis, Augusts Sukuts, Viktors Jansons
艺术家：Kristaps Gelzis, Reinis Suhanovs
主要承建人：MG buvnieks Ltd.
项目经理：Maris Gailis
结构工程师：Balts un Melns Ltd.
甲方：Society Zanis Lipke Memorial
用地面积：600m²
总面积：395m²
项目时间：2005~2009
建造时间：2008~2012
摄影师：©Ansis Starks (courtesy of the architect)

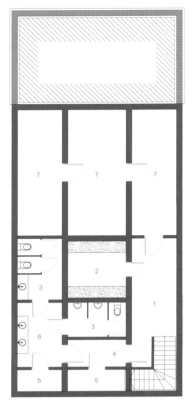

1 大厅 2 地堡 3 卫生间 4 走廊 5 杂物间
6 接待室 7 辅助间
1. hall 2. bunker 3. toilet 4. hallway
5. utility room 6. anteroom 7. auxiliary room

地下一层 first floor below ground

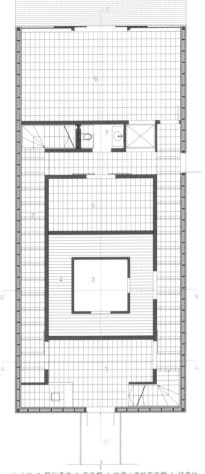

1 入口 2 展厅通道 3 苏克棚 4 地堡上面的苏克棚 5 信息台
6 休闲区 7 卫生间 8 室外露台
1. entrance 2. exhibition road 3. sukkah
4. sukkah above the bunker 5. information space
6. recreational area 7. toiltet 8. outdoor terrace

一层 first floor

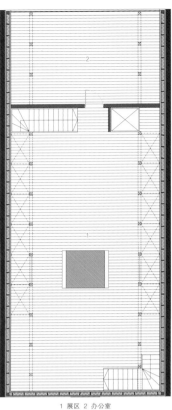

1 展区 2 办公室
1. exhibition space 2. office

二层 second floor

The territory of the museum is surrounded by solid black wooden fence. Passage through the enclosed tunnel that begins by the large entrance gate bears no suggestion of the real scale and structure of the building and it takes a while for the visitor to locate its center with the hidden bunker under the ground.

The visitors are emotionally directed along the pathways within the shed, then they ascend to the attic and arrive at the well through which one can see the pit in the basement, designed 3x3x3 meters wide with wooden plank beds resembling the room of the historic bunker. The attic is the main exhibition hall of the museum. Through the roof sparse sun beams enter the dusky room from the desired freedom outside. On the ground floor there is another detour around the bunker. A sukkah is built above the bunker – a symbolic Jewish divine temporary shelter from the cruel world, the desired home with painted transparent paper walls and small windows.

Construction materials: wood-frame building; overlapping wood plank roof and facade; wooden printed concrete walls; plank and wet polished concrete floors; plank ceilings. The construction is a passive house due to heat isolation system providing constant low temperature.

The building is funded by donations raised by Society "Zanis Lipke Memorial"

Zanis Lipke Memorial has been shortlisted in the category Culture at World Architecture Festival 2012, which took place in Singapore from October 3 to 5, 2012.

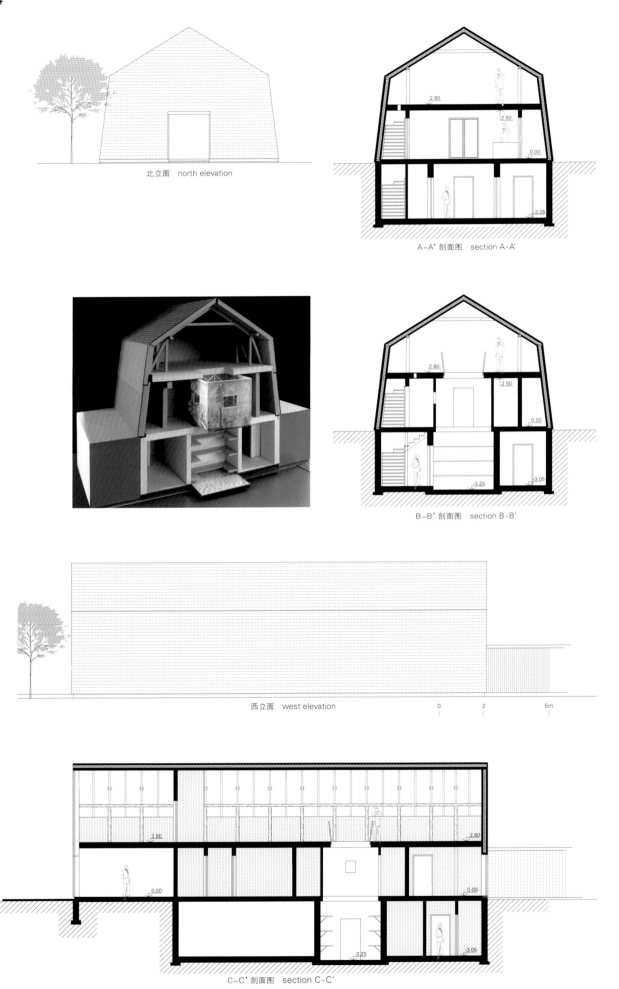

北立面 north elevation

A-A' 剖面图 section A-A'

B-B' 剖面图 section B-B'

西立面 west elevation

C-C' 剖面图 section C-C'

帕尔米拉纪念博物馆
WXCA

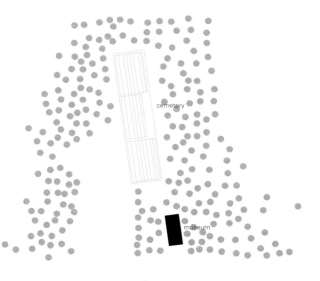

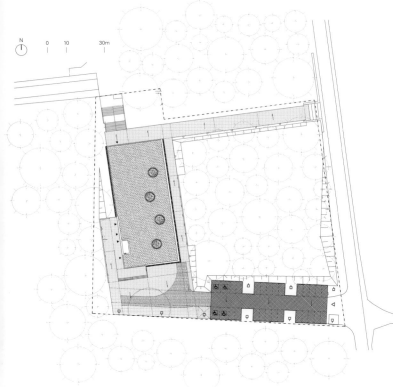

理念

帕尔米拉纪念博物馆是一片沿着围绕公墓的松树和桦树森林而建的博物馆区。这座建筑就是Kampinos国家公园的一部分,以玻璃和钢墙分隔开来,上面由绿色房顶覆盖。展览空间位于绿树丛中,用以见证过去的悲剧,建筑的苦行主义形式及严格的二手材料形成了展览内部的背景。

地理位置

博物馆毗邻一个普通的公墓,这座公墓是1939年至1941年纳粹统治时期的遇害者陵墓,它以Romuald Grutt and Ewa Śliwińska的项目为蓝本,建于1948年。建筑物的地理位置按照公墓的布局设计,保留了不规则形状的场地的合法要求。建筑位于人行道和南端的客人停车场之间,神圣和世俗的分界是通过沿着东立面设计的一条通道来完成,这条通道从停车场延伸至公墓,位于钢质博物馆的正面和混凝土阻力墙之间,呈斜坡状态。

形式及功能

建筑内墙和外墙形成了一个整体,二手材料是展览的背景,建筑不应干扰公墓内埋葬的遇害者的私人物品的感性信息。建筑内简单清晰地划分了不同功能。建筑入口位于公墓主巷的尽头,游客进入玻璃大厅空间内,右边是接待处和衣帽间,这里是用钢板封闭的小隔间,不限制内部体量和看向外部的视野,大厅尽头有培训室以及通向咖啡间的走廊、洗手间以及带舞台的办公区域。长方形的展览空间用钢筋混凝土墙壁分开,展厅利用开阔的玻璃来面对公墓,游客可以看到尽头的三个白色十字架。玻璃围合的院子栽种了松树,院子照亮了展览空间,将展览的特别部分分隔开来,并且对参观的方向进行了规划。

材料及建造

应用的材料符合场地及选址的重要性。选材严格,包括混凝土、玻璃、考顿钢,与Kampinos国家公园的环境以及公墓陵墓的混凝土十字架相搭配。

环境

由于其地理位置特殊,该建筑受到很多法律的限制,必须在人工制品保护、保存、环境及火灾保护上得到特别许可。外墙与森林线相隔小于12m,必须保证一旦发生森林火灾对使用者具有保护作用,同时在建筑内部发生火灾时有安全出口。在建筑立面为玻璃时这些危险是真实存在的。建筑立面使用了双层玻璃,里面为防火层,外层玻璃用于防止肆意破坏公物,展品的安全通过各种低电流系统保证,建筑入口有一个保安岗亭,控制游客进入。顶端的绿色屋顶上的绿地和国家公园管理处协商过,不破坏当地的生态系统。

安装解决方案

博物馆设备符合展览需要,华沙历史博物馆的使用者对展览大厅提出了要求。建筑内部的微气候必要按照保护史前古器物的条件进行。

Memorial Museum in Palmiry

Idea

Memorial Museum in Palmiry is a museum area complying with the pine-birch forest surrounding the cemetery. The museum building is a part of the Kampinos National Park, separated with glass and steel walls, covered with a green roof. The exhibition space lies among trees – the witness of past tragedies. The ascetic form of the building and the severity of the used materials form a background for the exhibition inside.

项目名称：Memorial Museum in Palmiry
地点：Palmiry, Poland
建筑师：Zbigniew Wronski, Szczepan Wronski, Wojciech Conder
结构工程师：Sławomir Kaszewski, Sławomir Pastuszka
展览设计师：Marek Mikulski
电脑绘图师：Rafał Kłos
用地面积：8,738m²　总面积：1,133m²
可用楼层面积：998.30m²
造价：PLN 8,000,000
设计时间：2009~2010　竣工时间：2012.10
摄影师：©Rafał Kłos (courtesy of the architect)

Location

The museum borders on an ecumenical cemetery – a mausoleum of the victims of the Nazi regime in the years 1939-1941, founded in 1948 according to the project of Romuald Gutt and Ewa Śliwińska. The building's location responds to the layout of the cemetery, sustaining the legal requirements of the irregularly shaped lot. The object is located between the pedestrian alley, and the guest parking lot on the south end. The separation of the sacrum and profane was achieved by designing a passage along the eastern facade, from the parking lot to the cemetery, between the steel museum front and the concrete resistance wall, holding the slope.

Form and Function

The interior and the exterior of the building form a whole. The used materials are a background for the exhibition. The architecture is not supposed to interfere with the emotional message of the personal items belonging to the victims buried at the cemetery. A clear simple division of functions has been suggested inside the building. The entrance to the building is located at the end of the main alley from the cemetery. The visitor enters the glazed space of the hall, with a reception and a cloakroom on the right, closed within a steel panel booth, so as not to restrict the inside cubature and the view from the outside. The end of the hall holds an education room and a corridor to the cafe, the restrooms, and the office area with a backstage. The rectangular exposition space is separated with a reinforced concrete wall. In the direction of the cemetery the exhibition space opens with a wide glassing, directing the eyes of the visitor to the three white crosses at the end. The glass patios include pines. The patios enlight the exhibition area, separating particular parts of the exposition, and organizing the direction of the visit.

Materials, Construction

The applied materials match the importance of the place and its location. The severity of the materials such as concrete, glass, the corten steel, complies with the surroundings of the Kampinos National Park, and the concrete rough crosses of the cemetery-mausoleum.

Conditions

Due to its location, the object was under numerous legal restrictions, it was necessary to get specific permits in the area of artifact protection, conservation, environment and fire protection. The distance from the outer walls to the forest line is smaller than 12 meters. It was necessary to ensure both fire protection for the users in case of a forest fire, and a secure exit in case of a fire inside the building. Such hazard was real with glass facades. A two-layered glass facade is used, with an inner fire-proof layer. Outer glassing is used for protection against vandalism. The security of the exhibits is ensured by various low-current systems. The entrance to the building holds a guard post, designed for access control. The greenery of the green roof has been discussed with the management of the National Park, not to disturb the local eco-system.

Installation Solutions

The museum facilities are suited for exhibition needs, the user of the Warsaw Museum of History has stated the requirements for the exposition hall. The micro-climate inside is determined by the necessity to ensure adequate conditions for the artifacts.

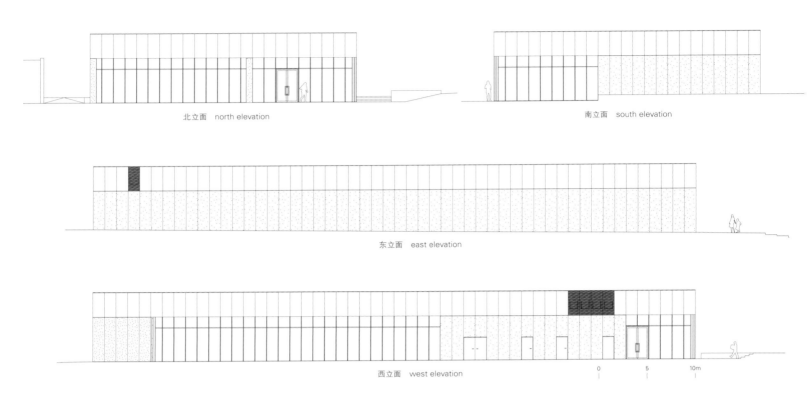

北立面 north elevation

南立面 south elevation

东立面 east elevation

西立面 west elevation

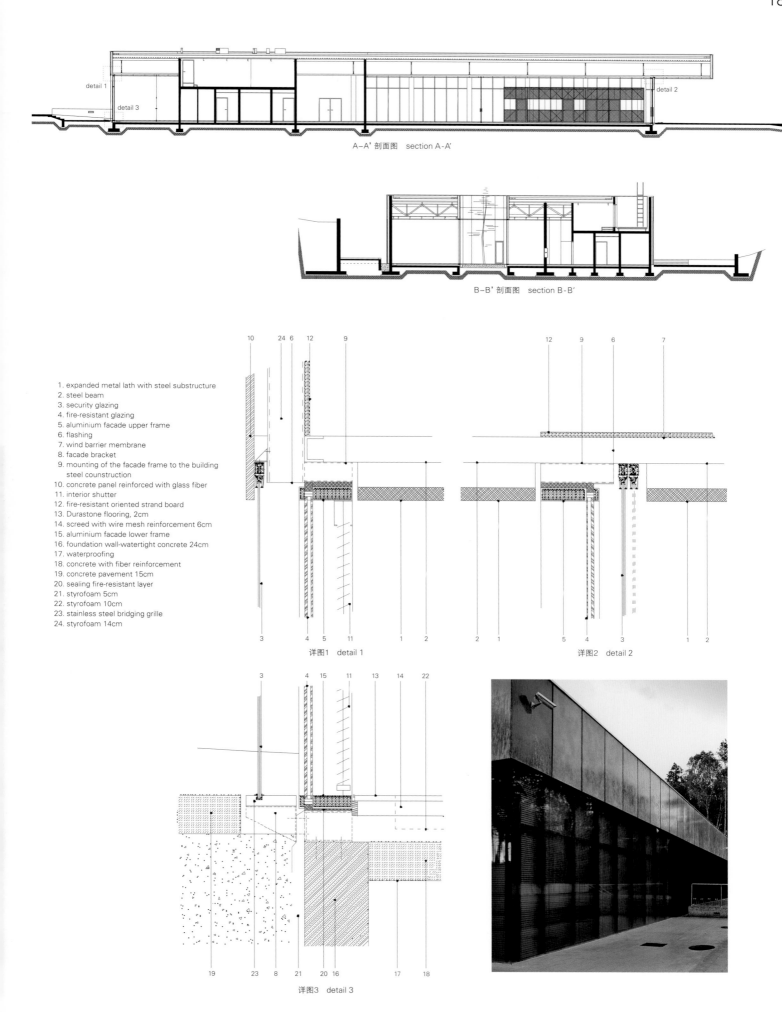

A-A' 剖面图 section A-A'

B-B' 剖面图 section B-B'

1. expanded metal lath with steel substructure
2. steel beam
3. security glazing
4. fire-resistant glazing
5. aluminium facade upper frame
6. flashing
7. wind barrier membrane
8. facade bracket
9. mounting of the facade frame to the building steel counstruction
10. concrete panel reinforced with glass fiber
11. interior shutter
12. fire-resistant oriented strand board
13. Durastone flooring, 2cm
14. screed with wire mesh reinforcement 6cm
15. aluminium facade lower frame
16. foundation wall-watertight concrete 24cm
17. waterproofing
18. concrete with fiber reinforcement
19. concrete pavement 15cm
20. sealing fire-resistant layer
21. styrofoam 5cm
22. styrofoam 10cm
23. stainless steel bridging grille
24. styrofoam 14cm

详图1 detail 1

详图2 detail 2

详图3 detail 3

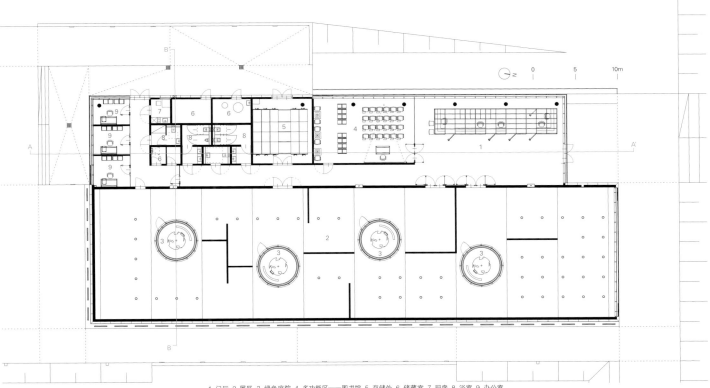

1 门厅 2 展区 3 绿色庭院 4 多功能区——图书馆 5 存储处 6 储藏室 7 厨房 8 浴室 9 办公室
1. foyer 2. exhibition 3. green patios 4. multifunctional zone – library 5. depository 6. store 7. kitchen 8. bathroom 9. office

一层　first floor

>>122
Wodiczko+Bonder
Is a Cambridge-based partnership established in 2003. Krzysztof Wodiczko [left] is an artist and professor of art, design, and the public domain at Harvard University in Cambridge. Julian Bonder [right] is an architect and professor at Roger Williams University in Bristol, Rhode Island. Focuses on art and design projects that engage public space and raise the issues of social memory, survival, and struggle and emancipation related to urban and domestic violence, war and post war trauma, immigration and global displacement, the Holocaust and genocides, the Desaparecidos in Argentina, the Civil War, and historical and present day slavery.

>>166
Zaigas Gailes Birojs
Was born in 1951 in Riga, Latvia. Graduated from Riga Technical University with a degree in architecture in 1975. Since 1991, she has been a leading architect of her own architecture company. Has received numerous national and international awards. Zanis Lipke Memorial has been shortlisted in the category of culture at World Architecture Festival 2012 which took place in Singapore from October 3rd to 5th, 2012. Is a founder and member of NGO Latvia Nostra-the regional organization of Europa Nostra and author of several books and publications.

© Ansis Starks

Cor & Asociados
Miguel Rodenas and Jesus Olivares have been running their own office, Cor & Asociados since 2006. They designed and built several public and private buildings in Spain, México, France and Ecuador. Received many awards for their work; Arquia Proxima(2008), Funeral Home in Pinoso (2009), and Kindergarden in Los Alcazares (2010). They also won the National Architecture Prize awarded by Ascer Ceramics in 2011.

awg Architecten
Bob Van Reeth received his diploma in architecture cum laude in 1968 from St. Luke's Higher Institute, Schaarbeek, Belgium. Is a self-employed architect since 1968. In 1972, he founded Architecten WerkGroep in Antwerp, Belgium with Geert Driesen, Filip Delanghe, Christine de Ruijter, Jan Verrelst and Ilse Van Berendoncks. From 1999 to 2005, he was appointed as the first Flemish government architect. In 2002, he reestablished Architecten WerkGroep as awg Architecten.

WXCA
Was founded in 2007 by Szczepan Wronski and Wojciech Conder in Warsaw, Poland. And is now led by Szczepan Wronski and Marta Sekulska-Wronska. The main focus of WXCA is to make this group knowledgeable, conscientious, and dynamic. In order to combine practice with a modern view, young designers co-operate with experienced architects in this office. Approach each project with passion and involve a lot of effort for them to be unique and functional at the same time.

Handel Architects
Michael Arad is a native of Israel. Was raised there, the U.K., the United States and Mexico. Came back to the United States and received a B.A. from Dartmouth College in 1994 and a master of architecture from the Georgia Institute of Technology in 1999. In 2006, was one of six recipients of the Young Architects Award of the American Institute of Architects. In 2012, was awarded the AIA Presidential Citation for his work on the National September 11 Memorial. In addition, was also honored in 2012 by the Lower Manhattan Cultural Council with the Liberty Award for Artistic Leadership.

Silvio Carta
Is an architect and critic based in Rotterdam. Lives and works in the Netherlands, Spain and Italy where he regularly writes reviews and critical essays about architecture and landscape for a diverse group of architecture magazines, newspapers and other media. In 2009, he founded the Critical Agency™ | Europe.

Nelson Mota
Graduated from the University of Coimbra in 1998 and received a master's degree in 2006 where he lectured from 2004 to 2009. Was awarded the Távora Prize in 2006 and wrote the book called A Arquitectura do Quotidiano, 2010. Is currently a researcher and guest lecturer at the TU Delft, in the Netherlands. Is a member of the editorial board of the academic journal Footprint and also one of the partners of Comoco Architects.

Karres en Brands Landscape Architects
Was founded in 1997 by Sylvia Karres and Bart Brands. Since then, they have been working on a wide variety of projects, studies and design competitions in the Netherlands and abroad. Their projects show the clear design strategy which tries to invent new ways of designing and planning. They want to make designs which have a strong identity, but flexible enough to be transformed. Believe that urban planners and landscape architects should almost work as "director" of design, planning, construction, and use of the site for a longer period not to dictate his own drawing board dreams, but to guide a process of interventions with the appropriate design tools.

>>84
Bayer & Strobel Architekten
Was founded by Gunther Bayer and Peter Strobel in 2006. They studied architecture and received an academic degree as graduate engineer in architecture from TU Kaiserslautern. The wohnhaus W that is one of their works is recognized with the BDA award in 2004 and Deubau prize in 2008. In addition, they were also awarded "best architects 13" and BDA award for Ingelheim Funeral Chapel in 2012.

tners
the chief architect and founder of G.Natkevicius & of the leading architectural design offices in Lithuania. among Lithuanian colleagues and gained the title of "Architect of the year". During the period of the company's existence, he received quite an amount of awards in the annual exhibitions of the Lithuanian Architects' Union and other important Lithuanian architecture exhibitions.

>>66
Giovanni Vaccarini
Was born in Italy in 1966. Graduated in 1993 with full marks in architecture from G.d' Annunzio University. Established Giovanni Vaccarini Architetto which is a design group of planning engaged in the preliminary, definitive planning, executive and direction works in the field of the urban planning and the architecture.

>>42
Zermani Associati Studio di Architettura
Paolo Zermani is the founder of Zermani Associati Studio di Architettura. Was an executive director of the international architecture magazine "Materia" from 1990 to 2000. Also has a significant production of projects published in several architectural magazines including A+U, Architectural Review, Domus, Deutsche Bauformen, etc. Has been invited to the Venice Architecture Biennale in 1991, 1992, 1996 and to the Milan Triennale in 1995 and 2003.

>>56
Architekt di Bernardo Bader
Bernardo Bader was born in Austria in 1974 and studied architecture at the University of Innsbruck. From 1998 to 1999, he worked for Feichtinger Architects in Paris and founded his own office in 2003. Since then, he was awarded more than 20 prizes. Now he is a member of Design Committee for intermediate water, Design Committee of Andelsbuch etc. Recently, he won 2013 Mies van der Rohe Award, held by European Union Prize for Contemporary Architecture.

>>72
Clavel Arquitectos
Manuel Clavel Rojo is the founder of the architectural design office, Clavel Arquitectos. Received more than 50 prizes for his architectural works. Taught at many universities and congresses of Madrid, Valencia, México DF, etc. and now he works as visiting professor at Miami University.

>>144
Gaeta-Springall Arquitectos
Was founded by Julio Gaeta[left] and Luby Springall[right] in 2001. Julio Gaeta has a PhD in architecture. Is a lecturer and invited professor in different schools and institutions of America and Europe. Was the editor of "Elarqa" (researching and publications) with more than one hundred publications of Architecture and Urbanism, both as author and editor. Luby Springall is an architect and plastic artist at the same time. Graduated from the University of Iberoamericana, Mexico and has post graduate diplomas in Art by the Royal College of Art, London, UK. For many years she has combined her work as an architect and artist. Her art work has been exhibited in London, Banff and Mexico. Teaches design at the University of Iberoamericana.

>>24
HGA
Joan Soranno[right], is an award-winning architect specializing in cultural and religious projects. Her passion for architecture is evident in her work style. With John Cook[left], she has created a design studio within the large-firm structure of HGA. The studio serves as an incubator for innovative design, in which her projects are community driven and technically challenging. John Cook, in collaboration with Joan Soranno, has worked on high-profile museum and cultural projects since joining HGA in 1997. As a project manager and project architect with considerable technical design skills, John has a proven track record working effectively with building committees, consultants and community stakeholders.

>>94
Andreas Fuhrimann Gabrielle Hächler Architekten
Andreas Fuhrimann[left] received an architectural degree from ETH Zürich. After graduation, he worked as design and planning architect for the architectural office Marbach + Rüegg.
Gabrielle Hächle[right] also received an architectural degree from ETH Zürich. After graduation, she was in assistant lectureship in the department of construction at ETH Zürich. Since 1995, she has been co-operating the architectural office with Andreas Fuhrimann. Both of them taught at ETH Zürich from 2009 to 2011 as guest professor and have been teaching at the University of Künste, Berlin since 2011.

C3, Issue 2013.5
All Rights Reserved. Authorized translation from the Korean-English language edition published by C3 Publishing Co., Seoul.

© 2013大连理工大学出版社
著作权合同登记06-2012年第145号

版权所有·侵权必究

图书在版编目(CIP)数据

终结的建筑 / 韩国C3出版公社编；于风军译. —
大连：大连理工大学出版社，2013.7
(C3建筑立场系列丛书；29)
ISBN 978-7-5611-8032-7

Ⅰ．①终… Ⅱ．①韩… ②于… Ⅲ．①丧葬建筑—建筑设计 Ⅳ．①TU251.6

中国版本图书馆CIP数据核字(2013)第162365号

出版发行：大连理工大学出版社
　　　　　（地址：大连市软件园路80号　邮编：116023）
印　　刷：北京雅昌彩色印刷有限公司
幅面尺寸：225mm×300mm
印　　张：11.75
出版时间：2013年7月第1版
印刷时间：2013年7月第1次印刷
出 版 人：金英伟
统　　筹：房　磊
责任编辑：张昕焱
封面设计：王志峰
责任校对：高　文

书　　号：ISBN 978-7-5611-8032-7
定　　价：228.00元

发　行：0411-84708842
传　真：0411-84701466
E-mail：12282980@qq.com
URL：http://www.dutp.cn